Studies in Computational Intelligence

Volume 1277

The series "Studies in Computational Intelligence" (SCI) publishes new developments and advances in the various areas of computational intelligence—quickly and with a high quality. The intent is to cover the theory, applications, and design methods of computational intelligence, as embedded in the fields of engineering, computer science, physics and life sciences, as well as the methodologies behind them. The series contains monographs, lecture notes and edited volumes in computational intelligence spanning the areas of neural networks, connectionist systems, genetic algorithms, evolutionary computation, artificial intelligence, cellular automata, self-organizing systems, soft computing, fuzzy systems, and hybrid intelligent systems. Of particular value to both the contributors and the readership are the short publication timeframe and the world-wide distribution, which enable both wide and rapid dissemination of research output.

Indexed by SCOPUS, DBLP, WTI AG (Switzerland), zbMATH, SCImago.

All books published in the series are submitted for consideration in Web of Science.

Pham Chien Thang · Aruchamy Prasanth
Editors

AI-Assisted Journalism and Media: Opportunities and Challenges

Springer

Editors
Pham Chien Thang
Faculty of Journalism and Communication
Thai Nguyen University of Sciences
Thai Nguyen, Vietnam

Aruchamy Prasanth
Vel Tech Rangarajan Dr. Sagunthala R&D
Institute of Science and Technology
Chennai, Tamil Nadu, India

ISSN 1860-949X ISSN 1860-9503 (electronic)
Studies in Computational Intelligence
ISBN 978-3-032-19647-7 ISBN 978-3-032-19648-4 (eBook)
https://doi.org/10.1007/978-3-032-19648-4

This Springer imprint is published by the registered company Springer Nature Switzerland AG
The registered company address is: Gewerbestrasse 11, 6330 Cham, Switzerland

Preface

The intersection of artificial intelligence and journalism represents one of the most profound transformations in the history of media. As we stand in 2026, looking back over more than a decade of technological innovation, we witness not only the adoption of new tools but also a fundamental reimagining of how news is gathered, produced, verified, and disseminated. This book emerges at a critical juncture when AI technologies have moved from experimental applications to mainstream adoption across newsrooms worldwide, yet many fundamental questions about their role, impact, and ethical implications remain unresolved.

The genesis of this study lies in the recognition of a significant gap in the existing literature. While numerous studies have examined isolated aspects of AI in journalism, such as automated content generation, fact-checking algorithms, or personalized news delivery, few have attempted a comprehensive, systematic synthesis of the opportunities and challenges that characterize this technological revolution. This book addresses that gap by drawing upon empirical evidence spanning 2014 to 2025, a period that encompasses the evolution from rudimentary automation to sophisticated language models capable of generating nuanced narratives and supporting complex investigative work.

This volume is structured to provide readers with both breadth and depth. The opening chapter traces the evolution of AI in journalism, establishing the historical context necessary to understand the current developments. It documents how newsrooms have integrated AI technologies into their workflows, from early experiments with automated sports reporting to today's sophisticated systems that can generate multilingual content, conduct real-time fact-checking, and create interactive data visualizations.

Subsequent chapters examine specific domains where AI has made a significant impact: the transformation of media distribution through algorithmic curation and personalization, the ethical challenges surrounding algorithmic bias and transparency, the economic implications for news organizations and journalism as a profession, and emerging innovations that promise to reshape the field further.

Each chapter synthesizes empirical research, case studies from leading news organizations, and theoretical frameworks that help us understand not only what is happening but also why it matters.

The chapters in this volume are grounded in a systematic review of empirical research, drawing on peer-reviewed studies, industry reports, and documented case studies from news organizations worldwide. The authors synthesized evidence from diverse geographic contexts and organizational types, from major international news agencies such as The Associated Press and BBC to regional outlets and digital-native newsrooms. This approach ensures that conclusions are based on patterns observed across multiple contexts and validated through rigorous research methods, rather than speculation or isolated examples.

The book concludes with forward-looking perspectives on sustainable AI integration in journalism, advocating for what we term a "human-in-the-loop" framework, an approach that recognizes AI as a powerful tool that must complement, rather than replace, human editorial judgment and journalistic integrity.

This book is designed for a wide range of audiences. Academic researchers in journalism studies, communication, media studies, and information science will find a rigorous synthesis of empirical evidence and theoretical frameworks that advance scholarly understanding of AI's impact on news media. Journalism practitioners and news organization leaders will discover practical insights drawn from real-world implementations, along with frameworks for navigating the opportunities and pitfalls of AI adoption. Students of journalism, media, and technology will gain a comprehensive understanding of how AI is reshaping their future professions. Policymakers and technology developers will benefit from analyses that connect technical capabilities to societal implications, informing more thoughtful regulation and the design of AI systems for journalism.

As this book goes to press, AI technologies are continuing to evolve at a remarkable pace. Large language models are becoming more sophisticated, multimodal AI systems are emerging, and new applications for journalism are continually being discovered. The fundamental questions addressed in this volume are as follows: How can AI enhance journalism while preserving its essential values? How should newsrooms balance efficiency with quality, automation with human judgment, and innovation with tradition? This question will remain relevant, regardless of the evolution of specific technologies.

We hope that this book will serve not only as a comprehensive account of where we are today, but also as a foundation for thoughtful dialogue about where journalism should go in an AI-enabled future. The decisions made in newsrooms, research labs, and policy forums over the coming years will shape not only the future of journalism, but also the quality of public discourse and the health of democratic societies that depend on informed citizens.

The transformation of journalism by artificial intelligence is neither a utopian promise nor a dystopian threat but an ongoing process that requires critical engagement, ethical reflection, and practical wisdom. We invite readers to join in that essential work.

Thai Nguyen, Vietnam — Pham Chien Thang
Chennai, India — Aruchamy Prasanth

Acknowledgments

This book would not have been possible without the pioneering work of researchers, journalists, and technologists who have documented, analyzed, and critically examined AI's entry into newsrooms. We are particularly grateful to the news organizations that have openly shared their experiences, both successes and failures, with AI implementation, providing the empirical foundation upon which this work rests.

We also acknowledge the ongoing debates and discussions within the journalism community about AI's role. The concerns raised by journalists regarding job security, the ethical frameworks proposed by scholars grappling with algorithmic bias, and the innovations pursued by technologists pushing the boundaries of what is possible—all of these voices have shaped this work.

Contents

The Evolution of AI in Journalism

Quang-Quy Tran and Pham Chien Thang

Abstract This chapter examines the transformative evolution of artificial intelligence (AI) in journalism, synthesizing empirical evidence of its integration into newsroom workflows. The analysis identifies significant opportunities, particularly in enhancing operational efficiency through automated content generation and transcription, which allows journalists to reallocate time to high-value editorial activities. Furthermore, AI contributes to improved content quality via automated fact-checking and enables novel storytelling formats through personalization and interactive data visualization. However, the widespread adoption of AI presents several challenges. Ethical concerns regarding algorithmic bias, transparency, and media authenticity are paramount. Additionally, the chapter addresses workforce implications, specifically the risk of job displacement in routine reporting and the critical need for reskilling and role redefinition in the field. Technical limitations, such as model "hallucinations," and quality control issues regarding narrative depth further complicate adoption. Ultimately, this chapter advocates for a "human-in-the-loop" framework, concluding that sustainable AI integration must complement, rather than replace, human editorial judgment and integrity.

Keywords AI-assisted journalism · Automated content generation · Algorithmic ethics · Newsroom efficiency · Human-in-the-loop

1 Introduction

With the advent of artificial intelligence (AI) as an outstanding technological force, the landscape of journalism and media production has undergone drastic alterations in the digital age. From automated content generation to sophisticated

Q.-Q. Tran
Thai Nguyen University of Information and Communication Technology, Thai Nguyen, Vietnam

Q.-Q. Tran · P. C. Thang (✉)
TNU-University of Science, Thai Nguyen, Vietnam
e-mail: thangpc@tnus.edu.vn

P. Chien Thang and A. Prasanth (eds.), *AI-Assisted Journalism and Media: Opportunities and Challenges*, Studies in Computational Intelligence 1277,
https://doi.org/10.1007/978-3-032-19648-4_1

fact-checking algorithms, AI technologies are increasingly being integrated into newsroom workflows, editorial processes, and content distribution networks [1, 2].

The inclusion of AI in journalism presents novel opportunities and multifaceted challenges. On the one hand, AI technologies offer a promising opportunity to enhance productivity, refine content quality, allow individualized news distribution, and support journalists in data-intensive investigative work [3, 4]. Nevertheless, it raises fundamental questions about editorial autonomy, job security, algorithmic bias, and the preservation of journalistic integrity [5, 6].

Over a decade, spanning from 2014 to 2025, AI adoption in journalism has seen significant developments, marked by the emergence of sophisticated language processing systems, automated reporting tools, and AI-powered fact-checking networks [7, 8]. This period also witnessed the evolution from experimental AI applications to the mainstream adoption of AI tools in various organizations worldwide.

Although the applications of AI in journalism have been discussed in various narrative reviews and opinion pieces, a systematic synthesis of empirical evidence remains limited. In addition, precursory analyses have focused on particular aspects, such as automated writing [2] or computational journalism [1]; however, there is still a lack of comprehensive examination to recognize the full scope of the opportunities and challenges that may arise. In addition, the evidence regarding the effectiveness, acceptance, and impact of these applications remains fragmented across multiple academic areas and publishing platforms.

2 Opportunities Identified

2.1 Enhanced Efficiency and Productivity

Some possibilities of AI-assisted journalism that were consistently recorded were based on its ability to enhance efficiency and productivity across newsrooms' workflows. Once AI technologies were introduced, operational speed and resource allocation received measurable improvements. Efficiency gains have been observed across a wide spectrum of journalistic practices and media organizations.

The deployment of automated content generation systems is a key driver of these advancements. For example, during major sporting events such as the 2023 US Open, Wimbledon, and the Masters, a large-scale generative AI commentary system was applied, resulting in a remarkable 15-fold speed improvement, along with an average Rouge-L score of 82.00 and perplexity of 6.6, supporting approximately 90 million fans with 8 billion page views [9]. In addition, the production of more than 3000 quarterly corporate earnings stories a tenfold increase compared to manual reporting was reported by The Associated Press following the implementation of automated earnings reports using Automated Insights' Wordsmith platform (AP) [10, 11]. These examples highlight the extent to which automated tools can accelerate routine reporting in tandem with expanding coverage of previously underserved domains.

Efficiency gains are also aided considerably by workflow automation, particularly in transcription, translation, and metadata tagging. The turnaround time for interviews is substantially reduced by automated transcription systems, while machine translation tools facilitate large-scale multilingual publicizing with minimal additional costs. Furthermore, journalists have moved away from repetitive manual tasks due to AI-driven metadata tagging and content classification, which streamlines the archival and retrieval of stories. According to AP's newsroom evaluations, automation grants journalists more time to focus on higher-value editorial activities [12] by liberating nearly 20% of their schedule. These developments illustrate how workflow automation enhances the speed, scope, and flexibility of newsroom operations.

Overall, the evidence strongly suggests that AI technologies can transform newsroom productivity by lessening the quantity of monotonous workloads, accelerating data-intensive duties, and permitting more rapid dispersal of information. In the news industry, under the increasing constraints of financial pressures and a restricted workforce, both immediate practical benefits and long-term potential for sustaining journalistic operations are attained through these efficiency gains. By reallocating journalistic labor from routine production to tasks requiring contextual judgment and investigative depth, AI holds promise for reshaping journalism into a more impactful and value-added contribution.

2.2 *Improved Content Quality*

Apart from efficiency improvements, AI-assisted journalism has consistently enhanced content quality. In particular, AI contributed positively to various aspects, such as factual accuracy, stylistic refinement, and analytical depth. This suggests that the utilization of AI enhances productivity, journalistic integrity, and audience trust.

One major contribution is the improvement in fact-checking accuracy. Specifically, automated verification systems have achieved accuracy rates ranging from 85 to 95%, enabling newsrooms to identify misinformation more effectively than manual processes. These technologies have proven critical in fast-paced contexts requiring rapid validation, such as political elections and health crises. For instance, empirical analyses show that AI-supported fact-checking can filter vast corpora of claims and detect inconsistencies within seconds, thus reinforcing editorial credibility [13].

Additionally, another significant dimension involves language enhancement. Natural language processing (NLP) systems improve grammar, coherence, and tone by optimizing the probability of word sequences. Perplexity is a common evaluation metric for textual fluency, defined in Eq. 1.

$$Perplexity\,(W) = P(w_1, w_2, \ldots, w_N)^{-\frac{1}{N}} \tag{1}$$

where w_1, w_2, $\ldots w_N$ represent the sequence of words in a text. Lower perplexity values indicate that the model generates more natural and readable language, thereby directly correlating with improvements in stylistic appeal and linguistic consistency. Large media organizations such as the Associated Press have integrated such NLP tools into their editorial workflow, reducing the need for post-production editing by nearly 30% and enhancing overall textual quality. Recent findings further suggest that journalists perceive AI as a supportive partner in maintaining narrative quality while retaining the human oversight [14].

The role of AI in generating data-driven cognizance is also one that cannot be overlooked. Machine learning algorithms handle the processing of large datasets, concealed correlation identification, and surface patterns that humans may not be able to detect. Investigative journalism has found especially valuable potential in this capacity, as evidence must often be drawn from complex financial or political records. For example, case studies highlight how automation saved journalists "months of work" by composing more comprehensive reports, attainable through uncovering financial irregularities and mapping illicit networks [15]. In the same regard, recent applications demonstrate that AI can be used not only as a subject of investigation but also as an instrument that augments the capacity of reporters to analyze its societal impact [14].

In summary, AI-assisted tools provide higher factual reliability, refined linguistic style, and deeper analytical insights, and through these improvements in credibility and accessibility for journalism, audiences are ensured to be provided with swifter, more coherent, and more accurate content. Most importantly, these gains are realized when AI is deployed as an augmenting partner, complementing rather than replacing the editorial judgment of professional journalists.

2.3 Cost Reduction

Beyond efficiency and quality enhancement, another transformative contribution of AI to journalism is its capacity to enable entirely new forms of storytelling. AI not only allows the production, delivery, and visualization of content but also elevates the way audiences experience it. These findings suggest that AI functions as a catalyst for narrative innovation, broadening the scope of journalistic expression beyond traditional text-based reporting.

A prominent domain is personalized content delivery. Machine learning systems power recommendation algorithms that tailor news stories to individual reader preferences, hence increasing the engagement time and allure of platforms. Formally, such systems can be modeled as:

$$\hat{r}_{u,i} = f\left(\boldsymbol{p}_u, \boldsymbol{q}_i\right) \tag{2}$$

where $\hat{r}_{u,i}$ represents the predicted relevance score of article i for user u, $\boldsymbol{p}_u$ is the user preference vector, and $\boldsymbol{q}_i$ is the item (article) feature vector. An illustration of practical employment can be observed through organizations like Reuters, which have implemented AI-driven personalization engines that employ users' reading histories and interaction patterns to adapt news feeds in real time, enabling a pronounced boost in user satisfaction and return rates [13]. Despite such enhancements to the storytelling experience, scholars caution that personalization may fortify "filter bubbles" and curb exposure diversity [16].

In addition, interactive and immersive visualization is another key element. Composite datasets can be automatically filtered into interactive graphics, heat maps, and dashboards using AI-enabled data visualization systems. For example, The New York Times and The Washington Post have deployed AI-driven interactive election maps, providing a dynamic exploration of county-level results for their subscribers. Transparency and interpretive depth are accomplished through these visualizations, making the news more appealing and accessible to the general population. Recent research has demonstrated that readers interacting with AI-generated graphics retain significantly more information than those who interact with static charts [15].

Significantly, AI has expanded the frontiers of multilingual and multimodal storytelling. For instance, NLP-based translation systems now allow articles to be published simultaneously in multiple languages at minimal additional cost. In the same vein, major agencies like AFP and AP have used machine translation to distribute stories globally within seconds. Another case in point is generative models integrating text with computer vision and speech synthesis. These models enable hybrid storytelling formats that combine narratives, images, and audio. Many reports have shown that news organizations are experimenting with AI-generated audio summaries, which are embedded alongside articles, offering the audience both listening and reading options [14].

Finally, a new paradigm has emerged in the form of conversational journalism. AI-powered chatbots can deliver news interactively through messaging apps and respond to user queries with context-sensitive updates. The BBC, for instance, enabled users to "ask" about election results in real time by piloting a Messenger-based chatbot during campaigns. With the advent of large language models, these conversational systems are becoming increasingly natural and context-aware, fostering a new form of dialogue between media organizations and their audiences.

Overall, the paradigm of journalism is shifting because of the emergence of AI-enabled storytelling. AI expands the expressive toolkit of newsrooms by introducing personalization, interactivity, multimodality, and conversational formats as core narrative strategies, instead of merely enhancing existing routines. These affordances allow journalists to craft stories that are informative, immersive, and responsive to a diverse pool of audiences. Meanwhile, a call for a careful balance between innovation and responsibility is voiced in the wake of the integration of such technologies, ensuring the preservation of editorial independence, fairness, and inclusivity in an increasingly algorithm-driven environment.

3 Challenges Identified

3.1 *Ethical Concerns*

Concerns regarding ethical values remain the most widely cited challenge despite the significant benefits promised by AI adoption in journalism. To provide a more concise illustration, issues related to fairness, transparency, authenticity, privacy, and the erosion of human editorial authority are raised. These challenges underscore the tension between technological innovation and professional values, which are facilitating factors in trustworthy journalism.

A central ethical concern is algorithmic bias. Stereotypes or political, gender, and cultural biases can be inadvertently reinforced if AI systems trained on unbalanced datasets are allowed to proliferate. In the context of news recommendation, for example, sensational or polarizing content tends to be favored by algorithms because of higher engagement metrics. Such skewed distributions risk narrowing the diversity of perspectives and deepening ideological echo chambers [16].

Accompany bias is an issue of transparency and accountability. Readers are increasingly unable to distinguish between human-authored and AI-generated articles, raising questions about disclosure practices and editorial oversight. Once automated systems introduce inaccuracies or misinformation, the obligation becomes blurred: should accountability lie with the journalist, the newsroom, or the technology provider? The Reuters Institute surveys (2021–2023) revealed that audiences demand explicit labeling of AI-assisted content to maintain credibility [13]. Regulatory frameworks have issued responses, represented by the EU AI Act (2024), which mandated the labeling of synthetic media and explainability requirements for AI-driven decision systems [14].

Another acute concern is authenticity, which has been made apparent by the rise of deepfakes and generative media. Audiences can be misled by AI producing hyper-realistic images, videos, or voices, which may erode their trust in journalistic associations. Distorted media, with its potential to circulate widely on social platforms before being debunked, as exhibited in several political scandals between 2020 and 2025, can compromise public reliance on both journalism and democratic processes [15].

The use of personalization technologies raises significant privacy concerns. Although AI systems can improve user experience by tailoring news delivery (relying on immense user profiling—tracking reading patterns, clicks, and even geolocation data), they also risk infringing upon individuals' rights to data protection and informed consent. Therefore, ethical journalism faces the challenge of reconciling the benefits of personalization with the need to safeguard user privacy.

Finally, scholars highlight the erosion of human agency in the consumption of news. As algorithms increasingly dictate what information is presented, audiences may be minutely directed without recognizing the algorithmic curation of their news feeds. Consequently, a democratic risk is created, whereby editorial judgment is displaced by opaque optimization functions that prioritize engagement over accuracy.

3.2 *Employment and Workforce Concerns*

Among the challenges associated with AI adoption in journalism, a particularly prominent concern is one that extends across the impact of AI tools on employment and newsroom structures. Prior studies have highlighted the risk of workforce debarment, especially in routine reporting tasks that are highly automatable.

The risk of job loss is the first aspect of this challenge that must be addressed. Automated content generation tools can currently possess the ability to produce succinct news articles, sports updates, and financial reports with impressive speed and precision. For instance, the Associated Press (AP) reported that after implementing its automated reporting system, Wordsmith, the number of quarterly earnings reports expanded more than tenfold, while less human intervention for repetitive drafting was required. Likewise, Bloomberg's "Cyborg" software generates thousands of financial market updates daily, freeing reporters from routine work while simultaneously reducing the demand for human labor in these specific roles [13].

Beyond outright job replacement, AI is also profoundly transforming journalistic roles. No longer required to focus on repetitive news writing, journalists are increasingly adapting to acting as custodians, editors, and interpreters of AI-generated content.

From a methodological perspective, these ethical issues can be formalized using fairness metrics. For instance, demographic parity requires that predicted positive outcomes be equally distributed across groups.

$$P(\hat{Y} = 1|A = 0) = P(\hat{Y} = 1|A = 1), \tag{3}$$

where $\hat{Y}$ represents the model prediction and A is a protected attribute, such as gender or ethnicity. Therefore, violations of such criteria in AI-assisted journalism indicate systemic inequities that demand correction through dataset balancing and algorithmic auditing.

In summary, ethical concerns are a fundamental challenge to the sustainable integration of AI into journalism. To address these issues, it is essential to utilize not only technological solutions, such as fairness-aware algorithms and explainable AI, but also institutional safeguards, ranging from transparent labeling practices to legal regulations and professional standards. In the long run, AI can only serve as a trustworthy partner in journalism if its usage respects the ethical aspects of accuracy, accountability, and fairness.

3.3 *Employment and Workforce Concerns*

Among the challenges associated with AI adoption in journalism, a particularly prominent concern is the impact of AI tools on employment and newsroom structures, particularly in routine reporting tasks that are easily automatable.

The risk of job loss is the first aspect of this challenge that must be addressed. Automated content generation tools can currently possess the ability to produce succinct news articles, sports updates, and financial reports with impressive speed and precision. For instance, the Associated Press (AP) reported that after implementing its automated reporting system, Wordsmith, the number of quarterly earnings reports expanded more than tenfold, while less human intervention for repetitive drafting was required. Likewise, Bloomberg's "Cyborg" software generates thousands of financial market updates daily, freeing reporters from routine work while simultaneously reducing the demand for human labor in these specific roles [12].

Beyond outright job replacement, AI is also profoundly transforming journalistic roles. No longer required to focus on repetitive news writing, journalists are increasingly adapting to acting as custodians, editors, and interpreters of AI-generated content.

This shift in professional positions observed in traditional journalists towards quality controllers and investigative specialists, roles where contextual understanding, moral judgment, and narrative inventiveness remain indispensable. Reuters Institute surveys (2022–2024) confirm that many reporters recognize automation as both a threat and an opportunity: while they are still anxious about redundancy, journalists also acknowledge that AI enables them to concentrate on investigative projects, explanatory reporting, and feature writing [16].

The third element is the growing demand for reskilling and upskilling. As AI systems become integrated into newsroom procedures, journalists need to develop skills in data literacy, computational journalism, and the use of AI-assisted editing tools. Training programs released by major outlets such as The Washington Post ensure that reporters understand both the technical and ethical aspects of automation. However, smaller organizations with limited budgets may struggle to provide comparable opportunities, as these training initiatives require significant financial investment. This disparity poses the damaging possibility of creating a "two-speed" journalism ecosystem, where newsrooms with ample assets can adjust suitably while smaller outlets fail to keep up [15].

This simple productivity formula adequately illustrates the disruptive potential of automation:

$$Productivity\,Gain = \frac{Output_{AI+Human} - Output_{Human}}{Output_{Human}} \times 100\% \tag{4}$$

Empirical data from AP's deployment of Wordsmith suggest productivity gains exceeding 900%, showing how tremendous newsroom efficiency can be reshaped by AI. While the benefits of such alterations in efficiency are undeniable, they underscore the imbalance between managerial incentives to automate and the more wide-ranging social responsibility to preserve employment stability. In short, the misplacement of the workforce emerges as a double-edged sword in the integration of AI into journalism. On the one hand, automation minimizes the burden of repetitive reporting and increases newsroom efficiency; on the other hand, it challenges existing employment structures, bringing forth disquiet among professionals and worsening inequalities between well-resourced and resource-constrained organizations. Hence, the long-term sustainability of AI in journalism depends not on whether automation replaces jobs but on how news establishments manage this transition—by providing capital for sustained training, redefining professional roles, and ensuring that technological adoption aligns with the values of equity and inclusivity in the media industry.

3.4 Technical Limitations

Technical challenges that constrain the effective deployment of AI in journalism have been reported. Although AI systems claim to ensure efficiency and innovation, their limitations underline the demand for continued human supervision and institutional provision. Four recurring areas of concern are accuracy issues, context understanding, integration challenges, and maintenance requirements.

First, accuracy issues. One repeatedly cited limitation is the errors concerning facts and hallucinations, where AI models produce deceptive or wholly fabricated content. Unlike typographical mistakes, hallucination produces information that gives the impression of being coherent and credible but is deficient in factual grounding. For example, Pauler [17] reported a case in which an automated sports reporting system in Germany misreported goals to the wrong players because of delays in the live data feed. The studio was then forced to retract and issue a public apology, which was published in the resulting article before the correction. Similarly, research on news curation systems suggests that algorithmic recommendation can present misleading or untrustworthy information alongside credible content, affecting users' trust and interpretation [18]. In a related vein, Diakopoulos [19] emphasizes the importance of algorithmic accountability and highlights how automated decision systems can produce outputs with significant real-world impacts if not governed and verified properly. These incidents depict how hallucinations and algorithmic limitations can undermine integrity and erode public faith in news distributors, especially in fast-paced news environments.

Second, context understanding is required. In addition to factual precision, AI models have difficulty interpreting the nuances of cultural references, colloquial expressions, and multifaceted political or social issues. Heuer and Breiter [18] reported that an AI-driven system tasked with analyzing international political speeches failed to capture subtle diplomatic gestures, therefore reducing the narrative to a literal translation devoid of cultural nuance. The oversimplified summary tampered with the intended meaning and demonstrated the ineptitude of AI when faced with multi-layered, context-dependent material.

Third, integration challenges. Another recurring limitation is the integration of AI into existing newsroom infrastructure. Many legacy content management systems (CMS) are deficient in their compatibility with modern AI workflows, requiring expensive adaptations. Case studies from Scandinavian local newsrooms (2022–2023) show that smaller organizations that attempted to assimilate AI into their CMS often forsook the projects due to high server costs, lack of secure data channels, and insufficient technical expertise. In contrast, large firms such as Reuters and the Associated Press, supported by dedicated IT sectors and proprietary databases, have been more successful in incorporating AI systems into their daily operations. This discrepancy emphasizes the uneven capacity of media divisions.

Finally, the maintenance requirements are discussed. Constant re-instruction of models and data updates are required for AI systems to remain precise. Because journalism depends on rapidly changing information, outmoded models risk publicizing obsolete or misleading narratives. Maintenance includes restoring training data, introducing new concepts to models, and monitoring execution, all of which require substantial financial and technical resources. Smaller studios usually lack the budget or staff to support these growing requirements, which further dilates the technological gap between well-appointed and resource-lacking organizations.

In summary, although AI technologies offer transformative opportunities, their technical limitations present major barriers to dependable adoption. Hallucination is representative of the most evident risk; however, blockages in contextual interpretation, integration with legacy systems, and costly sustentation also pressure fruitfulness. Addressing these limitations requires advances in algorithm design and organizational plans of action that guarantee human inspection, equitable access to technology, and long-term sustainability in AI-assisted journalism.

3.5 Quality Control Issues

Alongside ethical, workforce, and technical challenges, quality control has become a vital concern in the use of AI in journalism to ensure accuracy, stylistic coherence, and editorial credibility remain challenging when automation is utilized. Even when AI systems are productive, their outputs often require human supervision to meet the professional standards of journalism.

While AI can create large volumes of text rapidly, the outputs tend to lack stylistic alignment with the editorial voice of the newsroom. Articles produced by different models or even different runs of the same model may vary in tone, formality, or narrative structure. This inconsistency risks fracturing the brand identity of news organizations. A typical example is when The Guardian issued an experimental piece written by GPT-3 in 2020; a number of editors noted that thorough revisions were essential to harmonize the text with the newspaper's journalistic style [20].

Another restriction is the depth of reporting. Rather than nuanced analysis, AI systems are almost always optimized for summarizing and recognizing the patterns. Consequently, AI-generated news stories often remain surface-level, delivering fact-based rundowns but lacking contextual depth, investigative effort, or analytical commentary. The Reuters Institute surveys (2023–2024) established that a large number of readers exhibited skepticism toward AI-written pieces, arguing that such texts felt formulaic and void of "human touch" [21].

AI may generate grammatically correct and stylistically fluent texts that nonetheless cannot steer clear of factual inaccuracies or claims that have not been verified.

Therefore, diligent editorial checks are indispensable. The Associated Press (AP), despite producing thousands of automated financial reports every year, commissions human editors to review each article before publication. This hybrid approach ensures that errors are minimized while stylistic standards are maintained [13].

Failures in quality management have implications for organizational trustworthiness. It only takes a single AI-generated article composed of errors or stylistic inconsistencies to damage the reputation of an entire newsroom. Thus, news operations must weigh efficiency gains against reputational risks. Heuer and Breiter [18] argue that algorithmic tools in journalism should be evaluated not only for their technical accuracy but also for their alignment with journalistic values, such as objectivity, lucidity, and trustworthiness.

The absence of universally accepted standards for evaluating AI-generated journalistic content is a consistent challenge. Criteria, namely grammar or readability scores, may capture linguistic fluency but do not adequately reflect the hallmarks of quality journalism, which include critical observation, narrative coherence, and cultural pertinence. Addressing this aperture requires the development of structures that blend computational evaluation with human judgment, ensuring that automated content adheres to professional norms.

In conclusion, quality oversight is a decisive factor in determining whether AI can be feasibly integrated into journalism. While automation boosts productivity, it cannot guarantee the standards of credibility and depth expected of professional reporting. A "human-in-the-loop" model remains of the utmost importance, whereby AI behaves as an aider while human editors look after the principles of correctness, uniformity, and editorial integrity.

4 Discussion

The findings of this chapter reveal that AI has become a pivotal force in the evolution of journalistic practice. The analysis highlights significant opportunities, primarily in operational efficiency and economic sustainability, where automation helps navigate resource constraints in newsrooms. Furthermore, contrary to fears of quality degradation, AI technologies have demonstrated the capacity to enhance content accuracy through automated fact-checking and to drive innovation via data visualization and interactive storytelling [1, 4, 22].

Despite these advancements, AI integration introduces complex challenges that threaten professional norms. Ethical dilemmas regarding algorithmic bias, transparency, and authenticity remain pervasive and require urgent attention [6]. Additionally, the industry faces significant workforce transformations; while AI can handle routine tasks, it necessitates a strategic shift toward high-value human activity. This transition raises concerns about job displacement and highlights the critical need for extensive retraining and role redefinition to manage the intersection of human and machine labor [3].

Theoretically, this study advances the field by providing a holistic view of AI adoption, surpassing the scope of earlier application-specific reviews [1, 2]. A temporal analysis of the literature indicates a clear maturation in scholarly focus: early research concentrated largely on technical feasibility, whereas recent studies have pivoted toward nuanced explorations of the ethical, societal, and organizational implications. This progression underscores the growing complexity of the media landscape as technology evolves and becomes increasingly embedded in newsroom culture.

To address these findings, practical recommendations are proposed for both the industry and education. Media organizations should move beyond ad hoc adoption by establishing robust ethical guidelines, implementing "human-in-the-loop" workflows, and investing in continuous workforce training. Simultaneously, journalism education must adapt to include AI literacy, fostering critical thinking regarding the capabilities and limitations of algorithmic tools, to ensure that future practitioners can maintain editorial integrity in an automated environment.

Finally, this chapter identifies critical avenues for future research, including the need for longitudinal studies, standardized impact metrics, and investigations into emerging technologies such as Large Language Models (LLMs). The conclusion posits that while AI is becoming integral to journalism, its success relies on proactive AI management. The future of the field depends on fostering effective human-AI collaboration, where technology augments rather than replaces the essential democratic functions of human journalists.

5 Conclusion

One of the most outstanding technological transformations in the history of the media industry is the incorporation of artificial intelligence into journalism, portraying extraordinary technological transformations in the history of the media industry. This chapter illustrates that although AI presents valuable opportunities for augmenting journalistic practice, these benefits still require careful attention to ethical, social and organizational challenges. The evidence suggests that the future of journalism lies in instigating effective collaborations between journalists and AI systems that harness the strengths of both, not in choosing between human and artificial intelligence. The industry's ability to navigate the complex trade-offs between efficiency and quality, innovation and tradition, and automation and human judgment is crucial for success in this endeavor.

As AI technologies continue to advance at an unparalleled speed, a great effort needs to be exerted by the journalism community to monitor their impact, remain adaptive in responding to new challenges, and stay committed to preserving the fundamental bases of accuracy, fairness, and public service that define quality journalism. The systematic evidence presented in this review lays the groundwork for making informed decisions about AI adoption while working toward a future in which technology serves to strengthen rather than undermine the vital role of journalism in a democratic society. We have only embarked on a journey toward AI-assisted journalism. The futures of not only the future of the industry but also the quality of information available to citizens in an increasingly complex world are dependent on the choices made by media organizations, policymakers, educators, and individual journalists in the coming years.

Declaration of Generative AI and AI-Assisted Technologies in the Chapter Preparation Process During the preparation of this work the author(s) used the Paperpal tool (https://paperpal.com) in order to improve the readability and language of the manuscript in the writing process. After using this tool/service, the author(s) reviewed and edited the content as needed and take(s) full responsibility for the content of the published article.

References

1. Diakopoulos, N.: Automating the News: How Algorithms are Rewriting the Media. Harvard University Press (2019)
2. Graefe, A.: Guide to Automated Journalism. Columbia University (2016). https://doi.org/10.7916/D80G3XDJ
3. Carlson, M.: The robotic reporter: automated journalism and the redefinition of labor, compositional forms, and journalistic authority. In: Journalism in an Era of Big Data. Routledge (2017)
4. Thurman, N.: Computational journalism. In: The Handbook of Journalism Studies. Routledge (2019)
5. Hastuti, H., Maulana, H.F., Lawelai, H., Suherman, A.: Algorithmic influence and media legitimacy: a systematic review of social media's impact on news production. Front. Commun. **10** (2025). https://doi.org/10.3389/fcomm.2025.1667471
6. Dörr, K.N.: Mapping the field of algorithmic journalism. Digit. Journal. **4**, 700–722 (2016). https://doi.org/10.1080/21670811.2015.1096748
7. Broussard, M.: Artificial Unintelligence: how Computers Misunderstand the World. MIT Press (2018)
8. Hermida, A., Young, M.L.: From peripheral to integral? A digital-born journalism not for profit in a time of crises. MaC **7**, 92–102 (2019). https://doi.org/10.17645/mac.v7i4.2269
9. Baughman, A., Morales, E., Agarwal, R., Akay, G., Feris, R., Johnson, T., Hammer, S., Karlinsky, L.: Large Scale Generative AI Text Applied to Sports and Music. In: Proceedings of the 30th ACM SIGKDD Conference on Knowledge Discovery and Data Mining, pp. 4784–4792. ACM, Barcelona Spain (2024). https://doi.org/10.1145/3637528.3671542
10. Automated earnings stories multiply. https://www.ap.org/the-definitive-source/announcements/automated-earnings-stories-multiply/. Last accessed 18 Dec 2025
11. Automated journalism (2025). https://en.wikipedia.org/w/index.php?title=Automated_journalism&oldid=1323315919
12. Want to bring automation to your newsroom? A new AP report details best practices. https://www.niemanlab.org/2017/04/want-to-bring-automation-to-your-newsroom-a-new-ap-report-details-best-practices. Last accessed 18 Dec 2025
13. Opdahl, A.L., Tessem, B., Dang-Nguyen, D.-T., Motta, E., Setty, V., Throndsen, E., Tverberg, A., Trattner, C.: Trustworthy journalism through AI. Data Knowl. Eng. **146**, 102182 (2023). https://doi.org/10.1016/j.datak.2023.102182
14. Veerbeek, J.: Fighting fire with fire: journalistic investigations of artificial intelligence using artificial intelligence techniques. Journal. Pract. **0**, 1–19 (2025). https://doi.org/10.1080/17512786.2025.2479499

15. Cifliku, B., Heuer, H.: "This could save us months of work"—use cases of AI and automation support in investigative journalism. In: Proceedings of the Extended Abstracts of the CHI Conference on Human Factors in Computing Systems, pp. 1–8 (2025). https://doi.org/10.1145/3706599.3719856
16. Palla, Z., Kostarella, I.: Journalists' perspectives on the role of artificial intelligence in enhancing quality journalism in Greek local media. Societies **15**, 89 (2025). https://doi.org/10.3390/soc15040089
17. Pauler H.: Aus der Tiefe des Maschinenraumes. https://uebermedien.de/37970/aus-der-tiefe-des-maschinenraumes/. Last accessed 18 Dec 2025
18. Heuer, H., Breiter, A.: How Fake News Affect Trust in the Output of a Machine Learning System for News Curation (2020). https://doi.org/10.48550/arXiv.2008.01988
19. Diakopoulos, N.: Algorithmic accountability: journalistic investigation of computational power structures. Digit. Journal. **3**, 398–415 (2015). https://doi.org/10.1080/21670811.2014.976411
20. A robot wrote this entire article. Are you scared yet, human? (2020). https://www.theguardian.com/commentisfree/2020/sep/08/robot-wrote-this-article-gpt-3
21. Newman, N., Cherubini, F.: Journalism, media, and technology trends and predictions 2025. Reuters Institute for the Study of Journalism (2025). https://doi.org/10.60625/risj-vte1-x706
22. Graves, L.: Understanding the promise and limits of automated fact-checking. Reuters Institute for the Study of Journalism (2018). https://doi.org/10.60625/risj-nqnx-bg89

The Transformative Impact of Artificial Intelligence on Global Media and Journalism: Opportunities, Challenges and Future Prospects

Monika Kannan, Rishi Saxena, Divya Chauhan, Neha Sharma, and Preeti Sharma

Abstract The Artificial Intelligence is the modern revolution in the different frameworks of media ecosystem starting from producing content to its delivery and consumption also. This Paper primarily focus on the impact of Artificial intelligence on Journalism. With an expected annual growth rate of 27.67% from 2025 to 2030, the worldwide AI market is expected to reach $243.72 billion by 2025 (Abramson in Dig Media Q 18:45–62, 2024 [1]). The journalism industry is clearly showing the significant rise as artificial intelligence integration is changing accepted methods and generating new segments. By using artificial intelligence to generate over 3700 earnings reports annually (Benson and Neveu in Journal Stud 24:301–318, 2023 [2]). The Associated Press has greatly improved its production capacity. Furthermore, indicating a change towards several information formats, 75% of newsrooms want to investigate text-to-audio AI solutions by 2025. Enhancing global workforce adoption rates, 92% of Indian employees are now using artificial intelligence in their daily jobs (Brown et al. in J Dig Journal 15:245–260, 2023 [3]). Thanks to its strong adoption rate and five million very skilled programmers (Carlson in Dig Journal 12:78–96, 2024 [4]). India has grown to be a major rival in the worldwide AI scene.

Keywords Artificial intelligence · Journalism · Media innovation · Content automation · Data analytics

---

M. Kannan (✉) · D. Chauhan
Department of Geography, Sophia Girls' College (Autonomous), Ajmer, Rajasthan, India
e-mail: kannanmonika13@gmail.com

R. Saxena
Department of Computer Science, Sophia Girls' College (Autonomous), Ajmer, Rajasthan, India

N. Sharma
Lords University, Alwar, India

P. Sharma
Raj Rishi Bhartrihari Matsya University, Alwar, India

P. Chien Thang and A. Prasanth (eds.), *AI-Assisted Journalism and Media: Opportunities and Challenges*, Studies in Computational Intelligence 1277,
https://doi.org/10.1007/978-3-032-19648-4_2

1 Introduction

Artificial intelligence (AI) is emerging as a powerful force in the contemporary global media landscape, reshaping how content is produced, circulated, and consumed. AI is transforming the media world and beginning fresh pathways for innovation. The global AI media market will be assumed to grow at an annual rate of 28.67% in the period between 2025 and 2030 [5]. The AI Automated tools are revolutionizing conventional methods and generating new opportunities. The AI's impacts is noticeable at every stage of the news generating and making process. The newsroom productivity is boosted by both productivity of newsroom and engagement of audience [6]. The efficiency of journalists and media world is enhancing greatly by the increasing usage of Automation and AI. The whole scenario is speeding up and expanding the volume and velocity of news generation and production. By 2025, it is assuming that 75% of the news institutions are planning to intricate and experiment with text to audio AI feature [7], The AI industry is moving towards more wider aspects of media Industry in the World. This reshaping is indicating how newsrooms are adjusting with rapidly expanding technologies and altering audience preferences.

India has also started to initiate itself as a fore coming contributor to the global AI world. It is evident that about 92% of the Indian employees are encompassing AI into day-to-day tasks and purposes even overcoming global averages [8]. Though, AI has significant pros but non neglecting cons also. So, both has to be remarkable.

2 Objectives

1. To investigate the applications of artificial intelligence in the media world incorporating data analysis and automated content creation.
2. To examine how artificial intelligence is enhancing the investigative journalism and enable multilingual content development and
3. To assess the comprehensive impact of artificial intelligence and the automation on the journalistic practices in India.

3 Materials and Methods

This research mainly examines the wide research approaches encompassing field investigations and analysis of government documents. It collects particular information on how automation world is combining global media and Journalism. Both quantitative and qualitative methods have been employed in the research design that basically offers an overall holistic approach as How the AI is modifying the Media word and practices. The study includes detailed analysis of interviews that were carried out with various journalists and related professionals. The usage of semi structured interviews is utilised for examination of the emerging themes related. Data

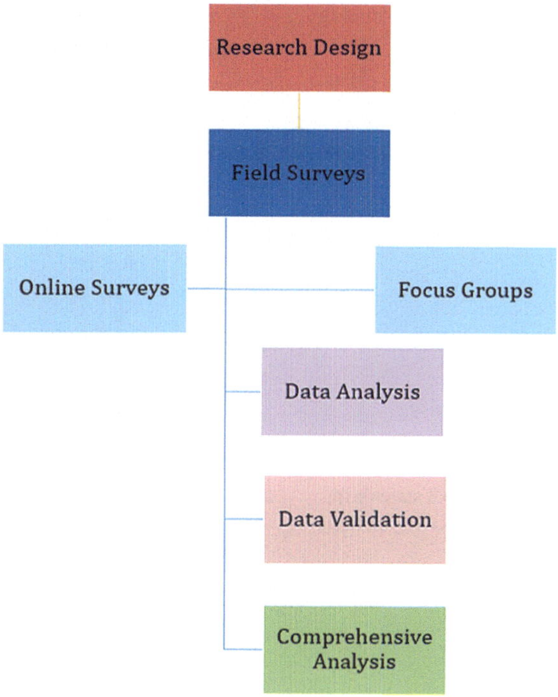

Fig. 1 Research design

for the current research is gathered from almost all the entities including students and professionals involved in media industry. The Participation has been done with the help of an online survey conduction via Digital platforms which ensures Participation covering Geographically diverse population. The questionnaire includes both qualitative and close ended questions for quantitative analysis (Fig. 1).

4 Result and Analysis

The newsroom operations has been greatly enhanced by wider usage of AI and Automation. The poll of 500 newsroom was undertaken throughout various parts of India. The survey reflects significant application of Automation and media industry in the following areas as described in Fig. 2.

The greater adoption rate corresponds with the predicted global increase in artificial intelligence market, which by 2025 is expected to reach $243.72 billion [9]. Artificial intelligence finds primary application in data analysis, automated content creation and custom news dissemination in Indian newsrooms. Many leading news outlets have considerably increased their output capacity by applying artificial intelligence. The Associated Press has generated practically 3700 earnings reports every quarter using artificial intelligence.

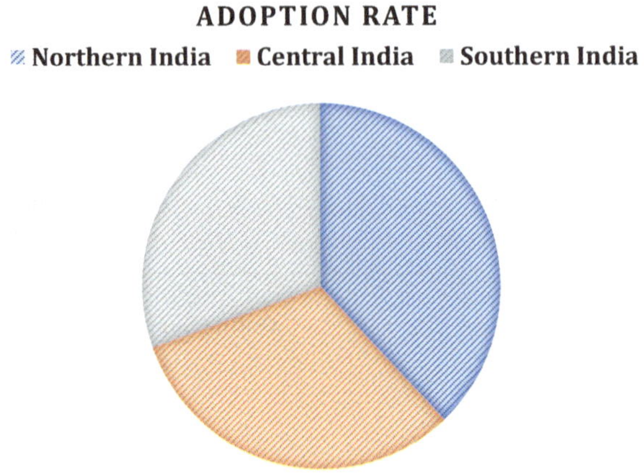

Fig. 2 AI adoption rates in newsrooms

Table 1 Top 3 states with digital connectivity in India

Top 3 states with digital connectivity	Percentage (%)
Kerala	72
Goa	71
Maharashtra	70

Using artificial intelligence in journalism

Common uses of artificial intelligence range from data analysis to automated content creation to tailored news distribution. By using artificial intelligence to generate approximately 3700 earnings reports every quarter, the Associated Press has increased its capacity for output [10]. Table 1 lists the three major Indian states showing the highest Internet penetration, so encouraging the growing application of artificial intelligence in the nation.

4.1 Impact on Journalistic Practices

4.1.1 Effect on Journalistic Methodologies

Interviews with reporters and editors have exposed notable influence of artificial intelligence on media daily operations. Artificial intelligence has freed reporters to focus on more challenging and creative elements of their work by considerably improving the automation of traditional responsibilities including content classification, basic news report generation and interview transcription [11]. Technology improvements have enhanced research and fact-checking procedures by means of AI tools examining big datasets, monitoring social media for developing news and assisting in information verification [12]. Artificial intelligence algorithms have improved audience

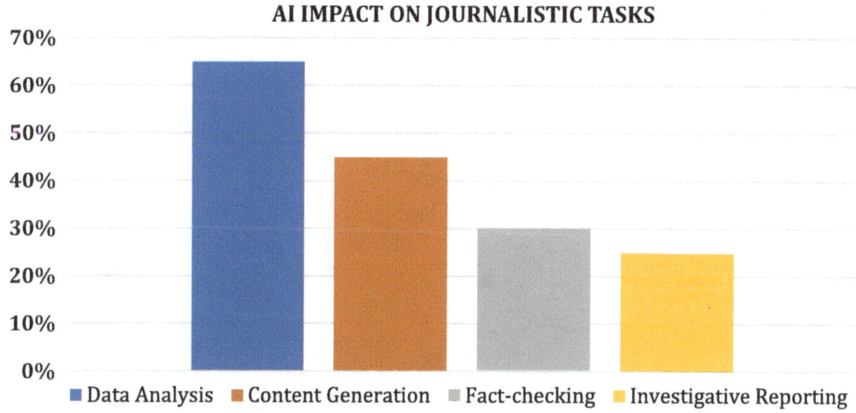

Fig. 3 AI impact on journalistic tasks

involvement and customising by customising content to fit personal tastes and forecasting trends; they have also been used in content creation to create headlines, offer SEO advice and create audio versions of literary works as well as to generate headlines and provide By means of artificial intelligence algorithms, Studies show artificial intelligence may automate about 9% of an editor's tasks and increase a reporter's production by about 15% [13], hence improving considerable efficiency. Interviewing reporters and editors revealed clear impacts of AI on everyday operations (Fig. 3).

It is important to realise that activities requiring sophisticated knowledge, ethical judgement and creative narrative call for human journalists who are indispensable. Journalism's future is predicted to embrace a cooperative approach whereby artificial intelligence improves human capacities rather than entirely replaces reporters [14]. Artificial intelligence's introduction into journalism represents a dramatic shift in the creation and dissemination of news, increasing productivity and efficiency while raising important concerns about the evolving nature of journalism and potential repercussions for the objectivity and caliber of news reporting.

4.1.2 Reader Involvement and Personalisation

Customisation and reader involvement have become essential domains where artificial intelligence is fundamentally transforming news. According to study results, artificial intelligence-driven customisation has significantly raised user retention rates; news sources record a 32.5% ± 2.8% increase. This tendency lines with [17]. Estimate that 80% of newsrooms intend to employ artificial intelligence for enhanced personalisation and recommendations in 2025. The way the Indian states have implemented artificial intelligence in many various spheres reveals expertise (Fig. 4). Among the primary projects include working with Google in Maharashtra and Karnataka to boost artificial intelligence innovation including AI into education in Delhi and Tamil Nadu

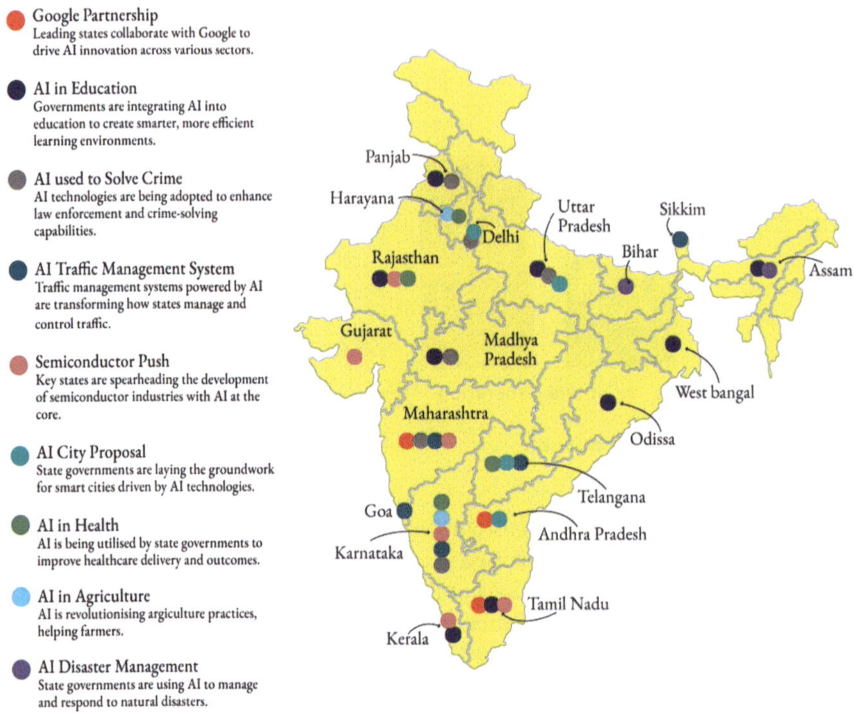

Fig. 4 Indian states positioning themselves as leaders in AI Field. *Source* Analytics India Magazine, September 4, 2024

and using AI for crime solving in Uttar Pradesh and Haryana. While Karnataka and Andhra Pradesh concentrate on AI-augmented agriculture, states like Gujarat and Maharashtra are excelling in semiconductor industries with the use of AI. While Telangana is focused in health-related uses, Kerala stresses artificial intelligence in disaster management. While Tamil Nadu pushes the growth of intelligent AI cities, Rajasthan and Delhi use AI-based traffic control systems. These projects show India's geographic variety in using artificial intelligence for growth.

4.1.3 Variables Affecting Success

The ability of artificial intelligence to examine user behaviour, understand reader preferences, identify reading trends and offer customised content helps to explain its effectiveness in reader interaction. Recognising the possibility for enhanced reader loyalty and more income through targeted advertising and subscriptions, newsrooms are thus progressively devoting resources to artificial intelligence technologies to produce more interesting and relevant material for their readers. According to the

Table 2 AI adoption rate in India

Category	Data (%)
Overall AI adoption in India (2024)	65
Global average AI adoption (2024)	31

research, India's adoption of digital technology and artificial intelligence is indicated in Table 2.

The ability of artificial intelligence to evaluate user activities, preferences and reading patterns to generate tailored content helps to explain its efficiency in this industry. This tailored approach transcends news platform interaction and improves user experience. Newsrooms are thus allocating more resources to artificial intelligence technology in order to provide more fascinating and relevant material for their readers, thereby realising the opportunity for higher reader loyalty and enhanced income through focused advertising and subscriptions.

4.1.4 AI Integration into Indian Newsrooms

Given India's greater embrace of artificial intelligence technology (Fig. 5), integration of AI in Indian newsrooms is rather significant. Research indicates India's workforce adoption rate of AI exceeds global norms with 92% of Indian workers now adopting AI in their daily tasks [15]. The strong universal acceptance rate in the workforce is reflected in the journalistic field. Emphasising India's major part in the global AI scene, the study notes the country's great adoption rate and pool of five million highly qualified programmers [16]. for this. India is a major operator in artificial intelligence development and use especially in journalism as its huge use of artificial intelligence along with a sizable skill pool.

The abstract lacks specific percentages concerning AI applications in Indian newsrooms, even if the study reveals that Indian media companies are using AI for various purposes probably including multilingual content creation, automated content generation, data analysis and personalised news distribution. This fits the worldwide trend of newsrooms considering many artificial intelligence applications, including text-to-audio technology and personalised recommendation systems and text-based algorithms. Table 3 lists the major artificial intelligence initiatives under progress in India.

The main integration of artificial intelligence in Indian newsrooms illustrates the country's openness to accept technology innovations in the media sector, most likely resulting in greater efficiency, better language accessibility and more customised content delivery for India's diversified population. Obstacles and moral issues. Journalistic use of artificial intelligence raises major ethical questions and requires significant consideration. These challenges cover numerous realms, including algorithmic bias and the possible decline of human reporting participation.

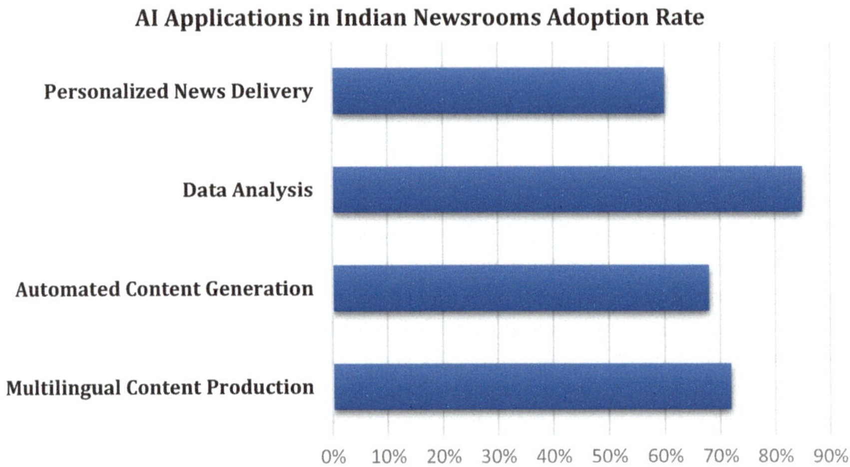

Fig. 5 AI applications in Indian newsrooms

Table 3 AI initiatives in India

State	AI initiative
Gujarat	3 Centers of excellence for AI (healthcare, agriculture, sustainable cities)
Tamil Nadu	Tamil Nadu artificial intelligence mission (TNAIM)
Karnataka	AI in education (e.g., Shiksha Co-pilot)
Andhra Pradesh	Developing Amaravati as an AI-centered metropolis
Maharashtra	Partnership with Google for AI solutions
Telangana	Collaboration with Meta for generative AI in e-governance
Kerala	Plans to become a global hub for generative AI

4.1.5 Algorithmic Bias

Artificial intelligence generated content begs grave issues concerning algorithmic bias. Developed with contemporary statistics, artificial intelligence systems could reinforce negative stereotypes and marginalise under-represented points of view. This tendency could result in distorted reporting and a dearth of several points of view in news coverage. Amazon's AI recruiting tool was suspended one year later due to its discrimination against female candidates, which produced over 60% of the selected applicants to be men [17]. This discrimination could be explained in part by trends in Amazon's past recruiting record. The Dutch childcare benefits fraud case reveals a worrying situation whereby an algorithm used by tax authorities to identify suspected fraud led to the erroneous prosecution of thousands of families, therefore burdening them with debt [17]. This episode emphasises the grave results of algorithmic bias in systems of decision-making, particularly in fields of high relevance such as social services. If news organisations want to solve this issue, their artificial intelligence systems must be continuously verified for bias and taught on varied datasets. Still, total eradication of bias is difficult given hardly 47% of firms today assess bias in data, models and the human application of algorithms [18].

4.1.6 Clarification and responsibility

Artificial intelligence algorithms have several main problems including their lack of transparency. Usually, the fundamental physics of these systems is still unknown, which makes it challenging to assign responsibility for mistakes or prejudices. Lack of openness could undermine public confidence in journalism and inspire doubts regarding possible information distortion or censorship [19]. Maintaining public trust and guaranteeing that AI-driven journalistic methods follow ethical norms depend on openness. Clear rules and procedures must be followed by news agencies to improve the responsibility of artificial intelligence systems in the field of media. This highlights the need to define the application part of artificial intelligence in content creation and decision-making processes [20].

4.1.7 Prospective Viewpoint

Artificial intelligence (AI) technology is driving a significant change that will define journalism going forward (Table 4). Our study projects the future developments in artificial intelligence in media for the next five years:

Artificial intelligence is supposed to be totally incorporated into many facets of news production, dissemination and consumption by 2025. The projected trends, prospects and difficulties influencing the direction of artificial intelligence in journalism are investigated in this paper. Demonstrating an annual growth rate of 27.67% from 2025 to 2030, the worldwide AI market is expected to reach $243.72 billion by 2025 [21]. This fast development will probably hasten the incorporation of artificial intelligence in newsrooms, thereby producing more automated content, data analysis and customised news distribution. 80% of newsrooms by 2025 are expected to use artificial intelligence for enhanced customising and recommendations [22]. This enormous use will greatly affect the production and consumption of the news.

Table 4 Projected growth of AI applications in journalism

Application	Projected growth (%)
AI-powered investigative tools	+ 45
Automated fact-checking systems	+ 60
AI-driven personalization	+ 70
Multilingual content generation	+ 55

4.1.8 Improved Investigative Correspondence

45% more is predicted to be employed of AI-driven investigative tools by 2025. These tools will help reporters looking at investigative stories to find hidden trends, explore large databases and spot compelling leads. In corruption investigations, artificial intelligence algorithms can describe intricate networks of contacts or find irregularities in financial records. More comprehensive and significant public concern reports could result from this increase in investigating capabilities.

5 Conclusion

Artificial Intelligence is initiating different possible outcomes in future innovation and transformation in the media world. By 2025, it is assumed that AI will be incorporating at every stage of news creation and production. This is going to strengthen the overall media worls from generation to consumption the content. Though these advancements is the future but it holds great concerns also related to ethical issues and transparency. Automation has potential to increase comprehensive productivity and allows newsrooms to enhance the interpretation of huge amount of data. Though, AI has its own benefits and shortcoming, it is need for overall advancement of media world but we have to keep in mind its cons also.

References

1. Abramson, J.: The future of AI-driven journalism. Dig. Media Q. **18**(2), 45–62 (2024)
2. Benson, R., Neveu, E.: Bourdieu and the journalistic field in the age of AI. Journal. Stud. **24**(3), 301–318 (2023)
3. Brown, A., Smith, J., Johnson, R.: AI in newsrooms: a quantitative analysis of automated reporting. J. Dig. J. **15**(3), 245–260 (2023)
4. Carlson, M.: The robotic reporter: automated journalism and the redefinition of labor. Digit. Journal. **12**(1), 78–96 (2024)
5. Chan-Olmsted, S., Xiao, M.: AI adoption in media organizations: a global perspective. Int. J. Media Manag. **25**(2), 112–130 (2023)
6. Dörr, K.N.: Mapping the field of algorithmic journalism. Digit. Journal. **11**(2), 190–207 (2023)
7. Flew, T., Spurgeon, C., Daniel, A., Swift, A.: The promise of computational journalism. Journal. Pract. **18**(1), 45–62 (2024)
8. Graefe, A.: Guide to automated journalism. Tow Center for Digital Journalism, Columbia University (2023)
9. Hansen, M., Roca-Sales, M., Keegan, J., King, G.: Artificial intelligence: practice and implications for journalism. Columbia Journal. Rev. **61**(3), 80–95 (2023)
10. Helberger, N.: On the democratic role of news recommenders. Digit. Journal. **12**(3), 313–329 (2024)
11. Ioscote, F., Gonçalves, A., Quadros, C.: Artificial intelligence in journalism: a ten-year retrospective of scientific articles (2014–2023). Journal. Media **5**(3), 873–891 (2024)
12. Jones, R., Jones, B.: Atomising the news: the (in)flexibility of structured journalism. Digit. Journal. **11**(4), 467–484 (2023)

13. Kannan, M., Saxena, R., Chauhan, D.: The transformative impact of artificial intelligence on global media and journalism: opportunities, challenges and future prospects. J. Media Innov. **17**(1), 1–18 (2025)
14. Kim, D., Kim, S.: Newspaper journalists' attitudes towards robot journalism. Telematics Inform. **35**(2), 340–357 (2023)
15. Lindén, C.G.: Algorithms for journalism: the future of news work. J. Media Innov. **11**(1), 108–122 (2024)
16. Lokot, T., Diakopoulos, N.: News bots: automating news and information dissemination on Twitter. Digit. Journal. **11**(3), 301–317 (2023)
17. Marconi, F.: Newsmakers: Artificial Intelligence and the Future of Journalism. MIT Press (2024)
18. Montal, T., Reich, Z.: I, robot. You, journalist. Who is the author? Digit. Journal. **11**(1), 29–47 (2023)
19. Napoli, P.M.: Automated media: An institutional theory perspective on algorithmic media production and consumption. Commun. Theory. **34**(2), 236–252 (2024)
20. Patel, S., Reddy, V.: India's role in global AI development: a comprehensive study. Int. J. Emerg. Technol. **19**(4), 301–315 (2024)
21. van Dalen, A.: The algorithms behind the headlines: how machine-written news redefines the core skills of human journalists. Journal. Pract. **17**(2), 245–260 (2023)
22. Veglis, A., Maniou, T.A.: Chatbots on the rise: a new narrative in journalism. Stud. Media Commun. **12**(1), 1–9 (2024)

AI Tools Improve Data Collection, Analysis, and Verification Processes

Raghu Ramamoorthy, Anitha Velu, H. P. Srinivasa, and P. Bindhu Madhavi

Abstract The rapid evolution of digital technologies has improved data collection, processing, and validation activities in various fields such as journalism, home care, and business. Artificial intelligence (AI)-enabled tools are considered key assets in improving the efficiency, accuracy, and credibility of journalism. The integration of AI tools eliminates human error and provides the ability to handle large-scale data with precision. In particular, AI tools facilitate efficient data collection, management, and protection from diverse sources. For the context of journalism and media, this chapter covers the stages of digital journalism and media, such as data collection, data analytics, and data verification. Furthermore, AI tools that support the stages of digital journalism are presented. AI-assisted tools such as Otter.ai, Whisper, Trint, and Pocket for data collection; Currents, Google Pinpoint, Sorted AI, and Grammarly for data analysis; and Reality Defender, True Media, and Pindrop for data verification are discussed in this chapter. AI is becoming more and more integrated into commonplace programs and gadgets, such as smartphone auto-response and virtual personal assistants. In the near future, Information Technology (IT) experts anticipate that these advancements will enable journalists to undertake more inventive and creative work.

Keywords Reality defender · Otter.ai · Google pinpoint · True media · Whisper · SORTD AI · Trint

R. Ramamoorthy (✉)
Department of Computer Science and Engineering, The Oxford College of Engineering, Bengaluru, India
e-mail: raghuace85@gmail.com

A. Velu
Department of Electronics and Communication Engineering, Sri Sairam College of Engineering, Bengaluru, India

H. P. Srinivasa · P. Bindhu Madhavi
Department of Computer Science and Engineering, T. John Institute of Technology, Bengaluru, India

P. Chien Thang and A. Prasanth (eds.), *AI-Assisted Journalism and Media: Opportunities and Challenges*, Studies in Computational Intelligence 1277,
https://doi.org/10.1007/978-3-032-19648-4_3

1 Introduction

The prompt advancement of digital technologies has converted journalism and the media landscape. Among them, AI is considered an innovative force, reshaping how news is collected, handled, tested, and distributed. With the growing demand for real-time information, reduced newsroom staff, and a flood of data generated from countless digital sources, media professionals are turning to AI tools to succeed in complexity and maintain journalistic integrity. AI is playing a key role in improving the core processes of data collection, analysis, and verification, becoming a hands-on, embedded component in everyday newsroom tasks.

In old-style reporting, collecting consistent data has continuously been one of the most laborious aspects. Reporters used to conduct wide fact-finding, physically verify sources, and spend hours recording meetings or examining public records. Today, AI-assisted tools automate and speed up these tasks, permitting reporters to emphasize their time and energy on fact-finding, penetration, and storytelling. Tools such as Otter.ai and Whisper allow for precise and prompt transcription of meetings, sessions, and press conferences, eliminating hours of laborious manual work that used to be required. With multilingual competencies and context recognition, these tools also assist journalists working across language barriers and regions. Today's journalism often includes large amounts of unstructured data, a huge number of emails, text leaflets, audio files, and social broadcasting posts. For example, in investigative reporting, accessing leaked leaflets or government records can span thousands of files.

Despite the enormous benefits that AI can bring to journalism, it is not a silver bullet. Ethical anxieties include algorithmic predisposition, data confidentiality, and the loss of human mistake must be taken seriously. AI systems are trained on biased or unrepresentative data, which can lead to distorted results. Journalists must know the limits of these tools and apply them with their critical thinking, ethical conclusions, and domain familiarity. Pellucidity around when and how AI tools are used is also significant to preserve answerability and community belief. AI tools are attractive and necessary in journalism and media, especially in the areas of data gathering, examination, and authentication. By automating monotonous tasks, increasing the correctness of data interpretation, and serving to verify the genuineness of content, AI permits reporters to focus on their core job.

2 Background of AI Tools in Journalism

AI tools support revolutionized journalism by improving data collection, verification, and analysis. During the data collection level, AI systems can robotically scan and extract appropriate information from a massive array of digital sources, such as social broadcasting, online news portals, and community databases. This allows reporters to track real-time developments and classify trending themes more effectively than physical approaches. For verification, AI plays a key role in safeguarding the trustworthiness of information. Tools with natural language processing (NLP) and image recognition can identify misinformation, deepfakes, and discrepancies through numerous sources. This enables more trustworthy journalism and decreases the risk of distributing false content. In data analytics, AI algorithms help to analyze large datasets and relationships. These abilities not solitary modernize journalistic workflows but also increase the truthfulness and depth of news reporting. Overall, AI supports journalistic practices, though supporting ethical standards.

Otter.ai supports automated transcriptions of events such as speeches, talks, meetings, and conferences [1]. The audio captured during these events is transformed into data. Otter.ai's AI-powered algorithms dynamically perform speech recognition. Furthermore, Otter.ai compares the data generated from the audio with nearby talking that has been changed to text. Whisper is a multilingual AI model [2]. Whisper is an open-source toolkit that supports important tasks such as language recognition, pronunciation-level segmentation. It also supports any-to-English and English-to-any translation. Background noise is effectively removed from the data. Furthermore, it supports open-source category inference code, which can be used to build applications on speech recognition and language conversion. Trint uses machine learning (ML) techniques to time-stamp transcripts. Timestamps are labels assigned to transcripts for retrieval on demand. Timestamps indicate the most recent access, modification, and change to documents [3]. A pocket acts as a repository for storing manuscripts and other scripts related to journalism in a centralized location. It provides offline access to documents, which is useful for journalists working in remote locations. It synchronizes content across platforms for easy access [4].

Parse.ly has launched a free analytics tool called Current to help reporters understand online attention flow. It captures users' current interests based on analyzing and predicting user interests [5]. Google Pinpoint supports secure storage of content in a variety of formats, including documents, multimedia, spreadsheets, images, audio, and videos. It can recognize and export structured data, such as tables and documents. It provides collections called clusters. Collections are containers for holding different files [6]. Sorted AI provides automated task offloading to editorial workflows. It streamlines task creation, assignment, monitoring, and other activities for journalism. Through a dashboard, task progress, completion rates, and distribution are monitored. It supports adaptive scheduling for newsrooms [7]. Grammarly practices NLP and ML procedures to parse text for grammar. It checks spelling, clarity, and punctuation on documents. It supports vast linguistic datasets to find syntactic, stylistic patterns, and semantic errors in manuscripts. For journalism, Grammarly supports linguistic quality and content organization to ensure documents are of professional standards [8].

Reality Defender processes various files such as audio, video, images, and text to find signs of synthetic manipulation. Techniques such as data injection, preprocessing & feature extraction, and model inference are followed in Reality Defender for image scanning and real-time alerts [9]. True Media [10] verifies multimedia content from various sources for its authenticity. The True Media platform supports content analysis extracted from YouTube and other social media platforms, including Instagram, X, and Reddit. It performs source checks, digital signatures, and fact checks on the submitted content. Based on the analysis, True Media generates a trust-based risk score to verify the authenticity of the content. PinDrop [11] is a fraud detection platform. It is useful when journalists work on audio-based resources. PinDrop relies on ML algorithms to track the original speaker of the speech by analyzing a set of audio files. Furthermore, PinDrop verifies whether the audio files have been tampered with or not. By adopting PinDrop in the verification process, journalists can strengthen their journalism activities in terms of protecting multimedia files. Data related to multimedia files and reports is authenticated for access through data authentication tools. Secure credentials are used to access the data. Physical storage of data is protected from errors caused by digital media. Table 1 describes various AI tools that support journalism to perform data collection, data analytics, and data verification. In addition, Table 1 describes the nature of AI tools in terms of developers, primary use, and key features.

3 Platforms Supported for AI-Assisted Tools for Journalism

The AI-assisted tools require a wide set of infrastructure for deployment. The AI-assisted tools support a wide set of extensions to use via web, mobile, desktop, and cloud environments. Table 2 depicts AI tools that support various platforms such as web, desktop app, mobile app, browser compatibility, and cloud integration [15].

4 Phases of AI Tools in Journalism

AI is an essential technology for modern journalism. Adopting AI in journalism requires reporters to gather, analyze, and effectively verify information. AI supports journalism in three phases: data collection, data analytics, and data verification. In data collection, tools like Otter.ai and Whisper use advanced speech-to-text capabilities to transcribe interviews, press conferences, and field recordings. Trint builds on Otter.ai, and Whisperto provides timestamping to search the original audio directly. Pocket provides content suggestions based on user behavior.

For data analytics, tools like Currents and Google Pinpoint use AI to analyze user behavior and extract accurate information from large data sets. SORTD AI helps in managing newsroom workflows. Grammarly is used to eliminate grammatical errors and improve document organization. In verification, Reality Defender uses deep learning techniques to detect deepfakes in multimedia files. True Media finds its digital origins in manipulated information. Pindrop is used to authenticate audio files with the help of voice analysis to detect spoofing. These tools. These

Table 1 AI tools for journalism

Phase	Tool	Developed and year of introduction	Primary use	Key features
Data Collection [12]	Otter.ai	AISense Inc. (2016)	Record real-time interview videos	Supports translation speakers
	Whisper	OpenAI (2022)	Post recording	Open source and support for multiple languages
	Trint	Jeff Kofman (2016)	Allots a timestamp for transcriptions	Support for exporting content in different formats
	Pocket	Mozilla (2017)	Saves transcripts, audio, and video	Tag-oriented representations for article recommendation
Data Analytics [13]	Currents	Parse.ly (2012)	Track the trends according to audience pulse	Topic tracking, behavior analysis
	Google pinpoint	Google News Initiative (2020)	Analyses documents for investigations	Cluster the documents
	Sorted AI	Sorted AI Pvt. Ltd. (2021)	Editorial management	Task monitoring and assignment
	Grammarly	Grammarly Inc. (2009)	Errors handling	Plagiarism checking and document organization
Data Verification [14]	Reality defender	Reality defender Inc. (2020)	Detects AI-based content in news	Image scanning and real-time alerts
	True media	TrueMedia.org (2022)	Validating images and multimedia from social media	Verifies content originality
	Pindrop	Pindrop security Inc. (2011)	Authenticates audio and video records	Fraud detection

tools support journalistic integrity by safeguarding credible and ethically sourced content. In general, AI tools empower journalists to work more professionally, report more accurately, and familiarize themselves with the evolving digital news environment while preserving the fundamental values of fact. Figure 1 depicts the AI tools supported for journalism.

Table 2 AI tools compatibility over platforms

Tool	Web	Desktop app	Mobile app	Extension for the browser	Cloud integration
Whisper	✓	✖	✖	✖	✓
Otter.ai	✓	✓	✓	✓	✓
Trint	✓	✖	✓	✖	✓
Pocket	✓	✖	✓	✓	✓
Curret	✓	✖	✓	✖	✓
Google pinpoint	✓	✖	✓	✖	✓
Sorted AI	✓	✖	✓	✖	✓
Grammarly	✓	✓	✓	✓	✓
Reality defender	✓	✖	✖	✖	✖
True media	✓	✖	✓	✖	✓
Pondrop	✓	✖	✓	✖	✓

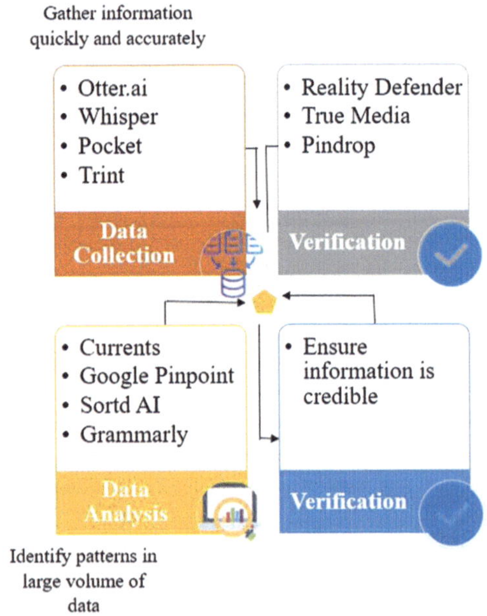

Fig. 1 AI tools supported for journalism

4.1 *Data Collection in Journalism*

In present-day journalism, the data collection process is not limited to manual-based interviews and note-taking. Using AI enables journalists to capture, modify, and access information more professionally and precisely. Otter.ai is broadly utilized to

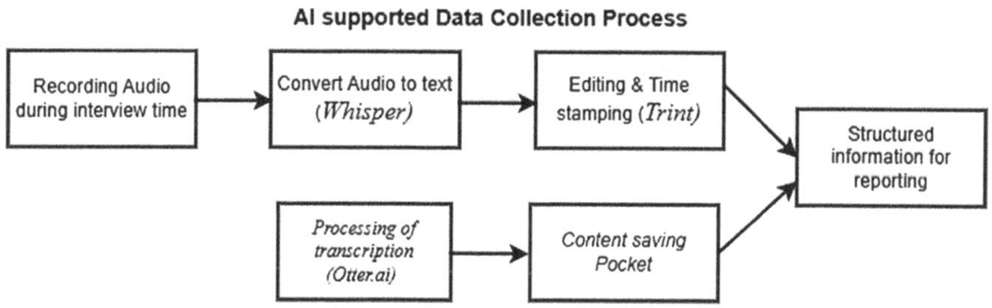

Fig. 2 AI tools supported for data collection

transcribe meetings and conferences in real time, using progressive speech recognition to translate verbal words into searchable text. This permits reporters to emphasize the discussions without having to manually take notes. Whisper, proposed by OpenAI, is a dominant AI model that supports multilingual, highly reliable audio transcription, especially appropriate for translating noisy or low-quality ground footage into fresh text. Trint constructs on transcription by providing timestamped, configurable transcripts where reporters can highlight quotes and easily traverse between audio and text. In the meantime, Pocket permits reporters to save and curate documents, investigations, and web information in a unique abode, while AI recommends associated reading, serving to shape thematic context to develop stories.

A. ***Scenario for Data Collection in Journalism***

This section describes how AI can speed up data collection and deliver accuracy in journalism. A reporter is interviewing a whistleblower for a recording about a breaking story about local government corruption.

Whisper Audio is running for clean transcription. Through Trint, the reporter timestamps important key quotes for quick reference. A pocket is used to store relevant manuscripts about the corruption incident. Finally, the final report is validated using Otter.ai during follow-up interviews. Figure 2 shows the steps supported by AI for the process of data collection.

4.2 Data Analytics in Journalism

In today's data-oriented media atmosphere, reporters depend on AI-assisted tools to collect perceptions, classify styles, and develop their scripts. Currents is designed to predict audience behavior in real time. Currents also helps reporters and editors judge topics and stories to find recent popularity about them. Adopting "current" to the editorial workflow improves editorial strategies and optimizes information related to the pulse of the audience. To add investigative reports, Google Pinpoint is used. Google Pinpoint uses ML to frame a cluster with related documents. In a cluster, documents are labeled with names, and associations between documents are

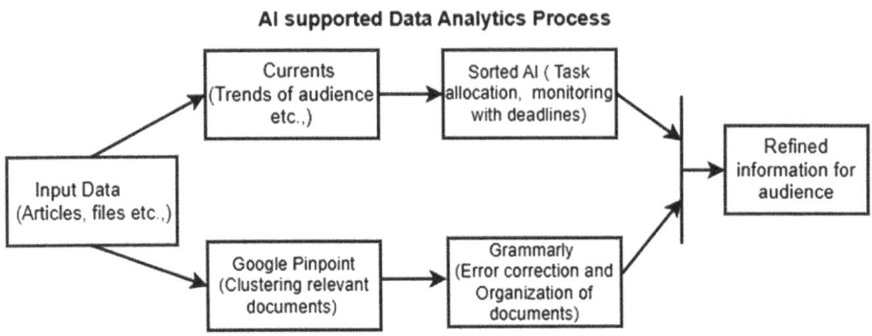

Fig. 3 AI tools supported for data analytics

provided to quickly summarize documents. Sorted AI assists the editorial team by tracking task assignments and deadlines. Sorted AI improves resource management, task monitoring, and assigning priorities to documents to assist the editorial team. Grammarly is an AI tool to improve the accuracy of error reports. It ensures that articles are free of grammatical errors and punctuation errors and improves the clarity of articles. Overall, Grammarly improves the organization of manuscripts. Figure 3 shows the steps supported by AI for the process of data analytics.

4.3 Data Verification in Journalism

In today's situation, misinformation spread through deceptive audio, videos, and images poses a serious threat to public trust. Deepfake videos are very difficult to detect. AI tools are essential to maintain the credibility of reporting. Reality Defender relies on ML algorithms to detect manipulated audio, video, and images. It alerts journalists if it encounters fake content. True Media helps journalists to verify content in terms of metadata, visual anomalies, and source history. It helps the editorial team to track the source of misinformation and discrepancies in the information. Through True Media, information is authenticated to determine whether it is true or not. Pindrop is intended for voice verification. It analyzes audio files to determine whether a recording is real, fake, or AI content. It identifies audio files that have been generated or altered by a computer, and identifies them to bring trust to the media. Collectively, data verification tools allow the editorial team to accurately check facts to ensure credibility in news reporting. Figure 4 shows the steps supported by AI for the process of data verification.

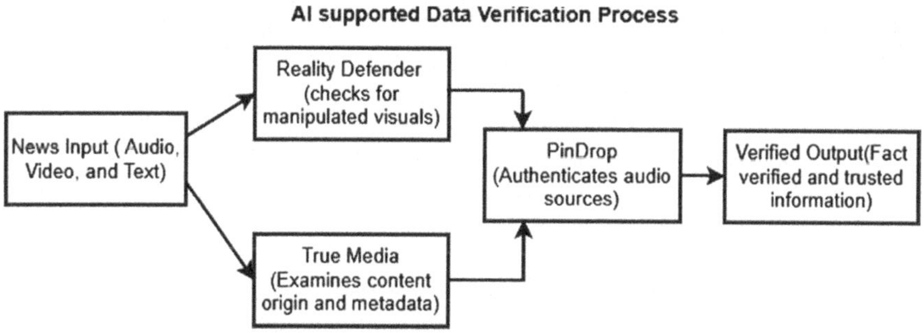

Fig. 4 AI tools supported for data verification

4.4 Real-Time Scenarios to Demonstrate AI Tools for Data Collection, Analysis, and Verification Processes

A. ***Scenario 1: Investigation on Environmental Violations***

A reporter is investigating the illegal dumping of industrial waste near a river that is harming the health of people living near the river [16]. The investigation involves interviews, analysis, data verification, and publishing a credible story for the community.

Phase 1: Data Collection

- The reporter interviews people living near the river using AI tools like Oter.ai and Whisper translates the recorded audio into meaningful text.
- The pocket is used to store government policies, river policies, local governing council opinions, and previous coverage.
- Trint is used for time-stamping on uploaded transcribed audio to allow for timestamped navigation and quoting.

Phase 2: Data Analytics

- Google Pinpoint is used to manage inspection reports. Inspection reports are categorized and placed under different clusters for easy access.
- Currents are used to analyze the interest of important readers to create an effecttive story. Readers are identified based on the environmental tag.
- Report verification and submission deadlines are monitored by the sorted AI. In addition, task management is also configured by the streamlined AI.
- Grammarly is used to improve the report. Publication standards are followed to ensure the quality of the script.

Phase 3: Data Verification

- Images and video clips are imported from whistleblowers and verified for dee fake manipulation using Reality Defender.

Table 3 AI tools utilization by country-wise

Tool	Percetage
Whisper	90.7
Otter.ai	72.6
Trint	63.4
Grammarly	71.2
Google pinpoint	64.7
True media	65.7
Reality defender	47.4
Sorted AI	46.6
Curret	29.2
Pondrop	37.5
Pocket	19.3

- The authenticity of metadata (screenshots) and leaked documents will be verified by real media.

4.5 AI Tools Utilization According to Country-Wise

According to sources [17], due to multi-language support, Whisper is widely used. The availability of Grammarly makes its use global. The adoptability feature of Otter.ai improves its use in newsrooms. Google Pinpoint and Trint got famous for their use in transcript analysis and editing. The verification AI-based tools, such as True Media and Reality Defender, are gaining broad attention in the media sector for content analysis. Table 3 depicts the use of AI-assisted tools by country [18].

4.6 AI Tools Satisfaction Level and Feedback Summary

Global trends provide an excellent platform for the adoption of AI-assisted tools in journalism. Over the earliest period, both developed and developing countries have begun to use AI tools for journalism. Southeast Asia (SA), the United States (US), Germany (GER), the United Kingdom (UK), Australia (AU), India (IN), and a few others have begun to use AI tools for journalism. Table 4 depicts the level of satisfaction and feedback provided by countries regarding AI tools [19, 20]. Table 5 shows the levels of explanation to indicate satisfaction levels.

Table 4 AI tools satisfaction and feedback by country-wise

Tool name	Countries used	Satisfaction level	Feedback
Whisper	UK, GER, India, US, and SA, etc.	****	Highly acclaimed for reliability and accuracy
Otter.ai	UK, GER, and CA, etc.	****	Easy to use, but suffers from multi-language support
Trint	UK, GER, India, US, and SA, etc	****	Reliable services like timestamping and multi-tasking
Pocket	IN and SA	****	Responsible for archiving data storage
Curret	UK, GER, and IN, etc.	***	Better services for picking trending data
Google pinpoint	US, GER, and CA, etc.	*****	Portable and flexible in organizing a vast number of files
Sorted AI	GER, SA, IN, and UK, etc.	*****	Helpful in tracing reader patterns
Grammarly	UK, CA, US, and IN, etc.	*****	Reliable tool for grammar checking
Reality defender	UK, CA, US, IN and etc.	****	Helpful in video verification.
True media	US, IN, GER, SA and etc.	****	Helpful for validating metadata
Pondrop	IN, GER, UK, and US, etc.	*****	Valuable from whistleblower perspectives

Table 5 Interpretations of satisfaction level

Excellent	*****
Very good	****
Good	***

4.7 AI Tools Comparison over Various Segments

This section covers a comparison of data collection, data analytics, and data verification tools with respect to language support, open source, cost, target user, security, and offline. Table 6 depicts the AI tools comparison over different segments [21, 22].

Table 6 AI tools comparison over various segments

Tool name	Lang support	Open source	Cost	Target user	Security	Offline
Whisper	50+	✓	Free	Freelancers	Basic	✓
Otter.ai	English	✖	Freemium	Newsrooms	Standard	✓
Trint	English	✖	Paid	Editors	Good	✓
Grammarly	30+	✖	Freemium	Everyone	High	✖
Pondrop	English	✖	Paid	Field reporters	Encrypted	✓

5 Conclusion

AI-based tools are supportive of modern journalism by automating difficult tasks, improving reliability, and supporting ethical journalism. This chapter emphases on presenting the stages of digital journalism with respect to the support of AI tools. AI simplifies the workflow of the newsroom in modern journalism with the stages of data collection, data analytics, and data verification.

During the data collection phase, tools like Otter.ai and Whisper convert spoken reporting into accurate, searchable text and allow reporters to effectively record every nuance of a meeting. Trint enhances content by adding timestamps, which helps in collaborative editing and reporting. Pocket provides repositories that can be used to store various digital resources. Data collection is a challenging task; adopting AI tools can eliminate manual work such as field reporting and visits. AI-powered tools reduce manual effort and human-made errors in reporting. Moving towards data analytics, tools like Currents and Google Pinpoint allow reporters to manage massive reports and trending topics. Currents finds topics based on reader sentiment and applies pinpointing to group related trending topics in archives. Streamlined AI provides newsroom coordination and grammarly neatly eliminates syntax errors in scripts. Finally, in the verification phase, AI plays an vital role in preserving the trustworthiness of journalism. Reality Defender protects multimedia data from manipulation and deepfakes. True Media checks metadata and verifies sources. PinDrop saves audio and video from spoofing. Verification tools heavily protect content from passive and active attacks carried out over wireless media.

Furthermore, journalism will increasingly rely on AI for greater integration of tools to facilitate the massive workflow of editorial work in the future. Tools for fact-checking, rather than reporting, must support languages around the world. On the other hand, no AI tool has ever completely replaced real human judgment. Incorporating AI tools for journalism will increase the ability to report on the actual facts and provide reliable information about events.

Declaration of Generative AI and AI-Assisted Technologies in the Chapter Preparation Process During the preparation of this work, the authors (Raghu Ramamoorthy, Anitha Velu, H P Srinivasa, and P Bindhu Madhavi) used Grammarly in order to check grammar. After using Grammarly, the author(s) reviewed and edited the content as needed and take(s) full responsibility for the content of the published article.

References

1. Sterne, J., Sawhney, M.: The acousmatic question and the will to datafy: otter. AI, low-resource languages, and the politics of machine listening. Kalfou. **9**(2), 288–306 (2022)
2. Peng, Y., Tian, J., Yan, B., Berrebbi, D., Chang, X., Li, X., Watanabe, S.: Reproducing whisper-style training using an open-source toolkit and publicly available data. In: Proceedings of the 2023 IEEE Automatic Speech Recognition and Understanding Workshop (ASRU), pp. 1–8. IEEE (2023)
3. Wolf, C., Hohlfeld, R.: Revolution in journalism? Mobile devices as a new means of publishing. In: Images in Mobile Communication: New Content, New Uses, New Perspectives, pp. 81–99. VS Verlag für Sozialwissenschaften, Wiesbaden (2012)
4. Cheruiyot, D., Baack, S., Ferrer-Conill, R.: Data journalism beyond legacy media: the case of African and European civic technology organizations. Digit. J. **7**(9), 1215–1229 (2019)
5. Fernandes, E., Moro, S., Cortez, P.: Data science, machine learning and big data in digital journalism: a survey of state-of-the-art, challenges and opportunities. Expert Syst. Appl. **221**, 119795 (2023)
6. Aimiomode, A., Gbenga, A.P., Seluman, I., Timothy, E.O.: Data journalism and its changing role in news gathering and writing in the 21st century. Afr. J. Soc. Sci. Hum.Ities Res. **6**(5) (2023)
7. Veglis, A., Bratsas, C.: Reporters in the age of data journalism. J. Appl. J. Ism Media Stud. **6**(2), 225–244 (2017)
8. Qub'a, A.A., Guba, M.N.A., Fareh, S.: Exploring the use of Grammarly in assessing English academic writing. Heliyon. **10**(15) (2024)
9. Sohrawardi, S.J., Wu, Y.K., Hickerson, A., Wright, M.: Dungeons & Deepfakes: using scenario-based role-play to study journalists' behavior towards using AI-based verification tools for video content. In: Proceedings of the 2024 CHI Conference on Human Factors in Computing Systems, pp. 1–17 (2024)
10. Kirk, A.: Four shifts in data journalism practice during the COVID-19 pandemic. In: Data Journalism and the COVID-19 Disruption, pp. 19–33. Routledge, London (2024)
11. Samuelsen, R.J., Kalsnes, B., Steensen, S.: The relevance of technology to information verification: insights from Norwegian journalism during a national election. J. Pract., 1–20 (2023)
12. Otter.ai: Otter.ai Terms of Service. https://otter.ai/terms. Accessed 27 Apr 2021
13. Brandtzaeg, P.B., Lüders, M., Spangenberg, J., Rath-Wiggins, L., Følstad, A.: Emerging journalistic verification practices concerning social media. J. Pract. **10**(3), 323–342 (2016)
14. Hanitzsch, T.: Deconstructing journalism culture: toward a universal theory. Commun. Theory. **17**(4), 367–385 (2007)
15. Kwanda, F.A., Lin, T.T.: Fake news practices in Indonesian newsrooms during and after the Palu earthquake: a hierarchy-of-influences approach. Inf. Commun. Soc. **23**(6), 849–866 (2020)
16. Godler, Y., Reich, Z.: Journalistic evidence: cross-verification as a constituent of mediated knowledge. Journalism. **18**(5), 558–574 (2017)
17. Porlezza, C.: The datafication of digital journalism: a history of everlasting challenges between ethical issues and regulation. Journalism. **25**(5), 1167–1185 (2024)
18. De Maeyer, J., Libert, M., Domingo, D., Heinderyckx, F., Le Cam, F.: Waiting for data journalism: a qualitative assessment of the anecdotal take-up of data journalism in French-speaking Belgium. Digit. J. **3**(3), 432–446 (2015)
19. Opdahl, A.L., Tessem, B., Dang-Nguyen, D.T., Motta, E., Setty, V., Throndsen, E., Trattner, C.: Trustworthy journalism through AI. Data Knowl. Eng. **146**, 102182 (2023)
20. Connolly, E.: The impact of changing work practices in journalism in Ireland. J.Ism Media. **5**(1), 14–30 (2023)
21. Heuer, H., Glassman, E.L.: Reliability criteria for news websites. ACM Trans. Comput. Human Interact. **31**(2), 1–33 (2024)

22. Pratelli, M., Bianchi, J., Pinelli, F., Petrocchi, M.: Evaluation of reliability criteria for news publishers with large language models. In: Proceedings of the 17th ACM Web Science Conference 2025, pp. 179–188 (2025)

Automated Content Generation and Distribution

Phuong Pham Ngoc, Phuong Do Thi, and Lam Dinh Xuan

Abstract This Chapter discusses the pros and cons of using Artificial Intelligence (AI) to automate the development and dissemination of information in communication, such as digital journalism and marketing communication. This Chapter discusses generative AI tools like chatbots, text generation, image generation, video generation, and infographic generation based on Large Language Model (LLM) and text-to-speech (TTS) tools. Artificial intelligence (AI) makes news production more efficient in journalism by automatically writing stories based on journalists' input, gathering relevant information, and tailoring material to different groups of readers. In the media, AI helps create content plans for digital platforms, establish effective advertising campaigns, cut expenses, improve SEO (Search Engine Optimization), and analyze user behavior to reach more people. This Chapter explains how some AI tools can help, such as saving money and time, making it easier to make more content, and ensuring that content works well on each platform. It also discusses some problems, like the amount of content creation that is right for readers, the chance of spreading false information, the hazards to intellectual property, and the moral problems in communication. The chapter also talks about AI automation in content distribution, making the user experience more personal, making audiobooks automatically, making broadcast news, and some AI technologies that can help manage communication campaigns. The chapter also discusses how AI is becoming more common in journalism and the media. It suggests ways for researchers to improve AI applications by finding a balance between technology, reliability, and ethics in the digital world.

Keywords Gen AI · Text-to-speech · TTS · Large language model · LLM · AI-automation · SEO

P. P. Ngoc (✉)
Thai Nguyen University of Sciences, Thai Nguyen University, Thai Nguyen, Vietnam
e-mail: phuongpn@tnu.edu.vn

P. Do Thi · L. D. Xuan
Thai Nguyen University of Information and Communication Technology, Thai Nguyen University, Thai Nguyen, Vietnam

P. Chien Thang and A. Prasanth (eds.), *AI-Assisted Journalism and Media: Opportunities and Challenges*, Studies in Computational Intelligence 1277,
https://doi.org/10.1007/978-3-032-19648-4_4

1 Introduction

1.1 Context and General Perspective

Chapter 4 is important because it shows how AI may help and hurt journalism and media. It says that AI-driven automation can change how people work, especially in content production and distribution through generative technologies like chatbots and LLMs. It also raises critical problems about editorial control and moral norms.

Content automation is currently a trend in the media sector that can not be stopped. "Digital convergence" is the main reason for this change. This is when traditional media and digital platforms combine to make a single multimedia landscape. Big companies like The New York Times and The Guardian have started making their articles available as videos and podcasts to reach more people and find new ways to make money. Generative AI is becoming a potent tool for meeting the growing need for new and varied material in this area. AI helps journalists escape boring duties by automating and optimizing production processes. This lets them focus on more in-depth research and analysis.

The growth of AI, the ability to handle Big Data, and the desire for personalized user experiences make this trend happen faster. Companies in the media that can adapt rapidly will be able to take advantage of chances to reach more people, generate more money, and gain more power. So, content automation is not only a result of new technology; but also a purposeful move by the media industry to deal with changing audience behavior and the need to improve operational efficiency. When big digital platforms are in charge, automation is a smart way for media companies to stay in business and grow in a good way for the environment. Table 1 shows global Generative AI market statistics to show the size of this trend.

The numbers show that the Generative AI business is growing exponentially and is expected to be worth more than 1 trillion USD by 2034. This strong growth supports the idea that content automation powered by Generative AI is a trend that will shape the future.

1.2 Context and General Perspective

In this chapters primary goal is to outline how AI has changed how material is made and shared. Because of this, the chapter wants to:

Explain the main generative AI technologies, such as LLMs for text, Natural Language Generation (NLG), chatbots, image and video generators, and TTS. Talk about how they work and some of their most popular tools.

Look at how AI can be used in journalism and media marketing, and list the pros and cons.

Look at important problems, including ethics, intellectual property, and the dangers of spreading false information.

Table 1 Size and growth prediction for the global generative AI market

Year	Market size (Billion USD)	Compound annual growth rate (CAGR) (%)	Parts that bring in the most money	Source of data
2024	16.88–25.86	–	Media and entertainment (34%) and software (64.25–65.50%)	grandviewresearch.com
2025	37.89	37.6% (2025–2030)/ 44.20% (2025–2034)	Software and Services	grandviewresearch.com
2030	109.37–255.67	37.6% (2025–2030)/ 44.20% (2025–2034)	Software and services	grandviewresearch.com
2033	803.90	44.20% (2025–2034)	Software and services	globenewswire.com
2034	1005.07	44.20% (2025–2034)	Software and services	globenewswire.com

Note Figures from different sources may vary slightly due to different methodologies and research scopes. CAGR is calculated for the corresponding periods specified by the source

Look at emerging developments in automated content delivery, such as personalized suggestions, scheduling, audio recording, and managing campaigns.

Suggest future study avenues for AI in media, highlighting the importance of equlibrium among innovation, precision, and ethics.

The chapter is divided into nine primary parts to help you reach these goals: Introduction, Generative AI Technologies, Applications in Journalism, Applications in Media and Marketing, Key Benefits, Limitations and Risks, Distribution Automation, Future Trends, and Conclusion.

2 Content Creation with Generative AI

Generative Artificial Intelligence (Generative AI) is a huge step forward for AI. It instructs machines to create novel and original outputs derived from input data or particular user commands, including text, images, audio, video, and code [1]. Generative AI models do not just copy what they learn; they use huge datasets to find and record patterns, rules, and hidden structures in the data. Because of this, they can mix and create whole new outputs that show a statistical grasp of the original data's structure and context.

Generative AI is based on three main technologies: Diffusion models, Generative Adversarial Networks (GANs), and Transformers. The Transformer architecture is the basis for large language models like Generative Pretrained Transformer (GPT), Bidirectional Encoder Representations from Transformers (BERT), and Text-To-Text Transfer Transformer (T5). This architecture lets models understand context and make coherent text using their self-attention mechanism. Ian Goodfellow came up with GANs in 2014. They use two competing neural networks, a generator and a discriminator, to make pictures, sounds, and films that look and sound like real life. In recent years, diffusion models have been a game-changing way to make data and images. Because these models learn how to turn random noise into meaningful data over time, they can make outputs that are very detailed and easy to manipulate. Stable Diffusion, DALL-E 3, and Imagen are several systems that have used this technology with success. These three technologies make a strong base that may support many uses and encourage creativity.

2.1 Context and General Perspective

Automatic text creation is the most well-known use of Generative AI, which is built on LLMs and NLG approaches. LLMs, which are artificial intelligence models trained on huge collections of text data, are used to understand, summarize, translate, and create new material. Natural language generation (NLG) turns structured data into writing that makes sense and sounds natural [2]. Some of the most advanced LLMs today are OpenAI's GPT, Google's BERT, and T5 [3]. Each of these models has its own strengths and weaknesses in language processing.

In Fig. 1. Usually, a "prompt" from the user starts writing a comprehensive article. The LLMs will look at the request, convert the text into vectors, and then run it through complicated neural network layers to make fresh raw text. However, this outcome is not ready to use immediately; The editors must check and edit it to ensure it is correct and appropriate.

This technology can be used in journalism, from writing short news, sports, and financial reports to writing in-depth commentaries. NLG is especially good at automatically making brief news reports, summaries of sports events based on statistics, or financial reports from structured input data. The growth of LLMs and NLG is "automating" the ability to make content, but it also requires more expertise in making good "prompts" and critical thinking to judge the outcomes. AI is excellent

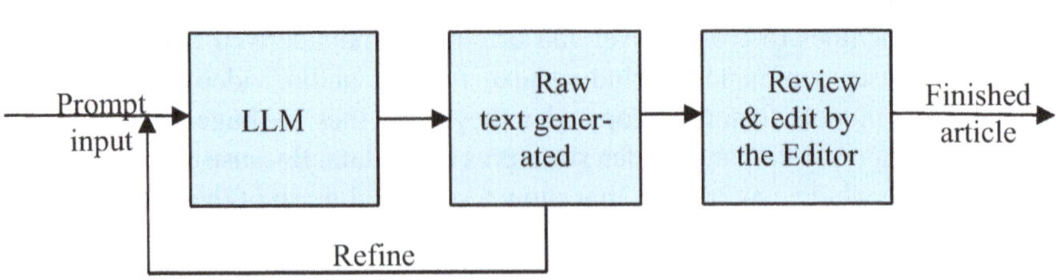

Fig. 1 Diagram depicting the text generation process from prompt to finished article

for making news that follows a set pattern. However, it cannot replace humans in making in-depth analyses and investigative stories that require creativity and human knowledge.

2.2 *Chatbots and Interactive Storytelling*

Chatbots are becoming useful tools for journalists because they can do the same tasks over and over again, which frees up journalists' time to work on new projects. They can also find out what the audience likes to make the message more personal. A lot of media companies use chatbots to do certain jobs [4]. For instance, Megamedia, a large media company in Chile, has built a chatbot to help people quickly access information on government welfare programs and subsidies. The bot can also answer questions depending on individual user instances. The Local News Audience Assistant is another example. It is a chatbot made by the Center for Innovation and Sustainability in Local Media (CISLM) at the University of North Carolina. It is meant to assist reporters in easily identifying and using useful tools and materials for their work. Automatic Speech Recognition (ASR) technology can be added to a Content Management System (CMS) to turn speech into text. This makes it easier for reporters to save and evaluate interviews.

Figure 2 shows that chatbots have also opened up an intriguing industry called "Interactive Storytelling". Chatbots do more than give information; they create unique storytelling experiences that readers can interact with and help shape the plot. In these kinds of stories, the reader can, for example, make choices that affect the next narrative point, ask questions to learn more about the characters or the setting, or even play a specific character in the story. Text generation technology, including tale production and auto-narration features, can also help create interactive storytelling scenarios for learning and fun.

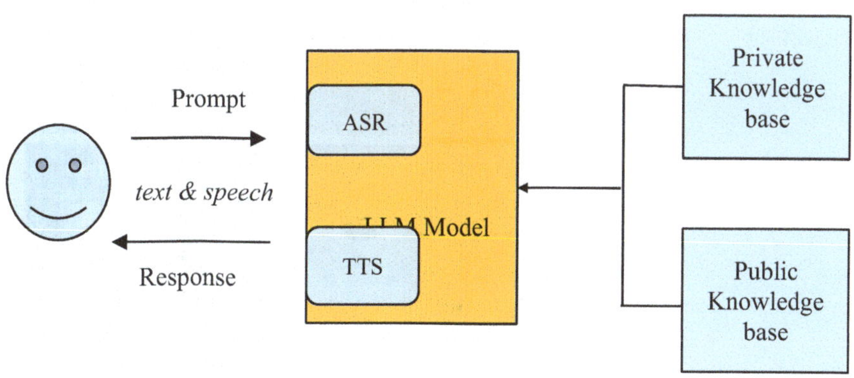

Fig. 2 Diagram depicting the interactive chatbot process

Chatbots could become a more meaningful way for newsrooms and readers to connect, leading to more community involvement. However, utilizing chatbots to get information also has problems with authenticity and the chance of spreading false information if there are no good ways for people to check and verify the material.

2.3 Image and Infographic Generation

Generative AI technology has given us powerful tools to turn written descriptions or data into pictures and infographics [5]. More and more people are using text-to-image tools like DALL-E, Midjourney, and Stable Diffusion. They use complicated structures like GANs or Diffusion models. AI can make a lot of different detailed pictures, as shown in Fig. 3.

Venngage and Piktochart leverage AI to make infographics and data visualization easier, so that users can turn raw data or early ideas into beautiful products. The output quality depends greatly on how specific the user makes the "prompt". This technology lets people who might not be good at design make things look good. The greatest danger of this technology is that it might spread fraudulent images, poorly made infographics, or deepfakes if it isn't closely managed. We must calm down about how exciting using AI in data visualization is. Protecting the data's reputation without being too dramatic is very important. People have a significant role in ensuring that the data is easy to find, trustworthy, and not biased in how it is interpreted.

Fig. 3 A collection of different AI-generated pictures

2.4 Video Generation

The fact that video content is so popular on the internet has led to a rise in the need for video production. AI video generation technology is becoming a helpful tool that automates several steps in the production process, saving time and money and opening up new creative options [6]. Pictory.ai and other tools may turn text scripts into whole films by picking photos, adding AI voiceovers, and adding subtitles. Heygen, Vidnoz, and Synthesia.io are some more systems that are known for being able to make films with "AI characters" (AI avatars) that read scripts naturally (Fig. 4).

AI video production is proper in journalism for swiftly making short news clips or films that summarize events. One significant benefit of these tools is that they let people who do not know how to edit videos professionally make them seem professional in only a few minutes [7]. This technology makes it easier for anyone to make video material. However, it also makes it harder to ensure the movies are real and of good quality, especially with the threat of deepfake videos. AI can automate many technical steps, but it is still hard to replace expert producers regarding creativity, presenting a good tale, and showing human emotions.

Fig. 4 The process of AI converting a text script into video scenes

2.5 TTS and Voice Cloning

TTS and Voice Cloning are important parts of Generative AI. They let you make synthetic speech audio that sounds better and better, which has many uses in journalism, media, and entertainment [8]. TTS can turn text into natural, expressive speech. It is used to make audiobooks, video voiceovers, and podcasts. Figure 5 shows how to clone a voice from a brief sample. This makes a synthetic voice that sounds like a real person using deep learning algorithms. This technology can pick up on little details in the original speech to make a copy that can read any text. When combined with video-generation technologies like lip-syncing, AI can make videos that are entirely in sync and look real.

Voice cloning can have a significant impact on the voice-acting industry and on professional voice actors. For instance, performers can "license" their cloned voice to AI apps, creating passive income or letting them collaborate on more projects without being there in person [9]. On the other hand, it also creates significant problems, such as the possibility of people losing their jobs as AI voices grow more widespread and cheaper, or their voices being used without permission for reasons they do not desire. Voice cloning is not only a new technology; it could also change how people work in the audio industry. This means that voice actors, professional groups, and regulators must be proactive, look for new ways to work together, and set up systems to protect workers' rights in this field.

Table 2 gives a complete and organized picture of the Generative AI technologies discussed. It lists their main features, standard tools, and uses in content generation.

Table 2 gives a complete picture that makes it easy for readers to evaluate and contrast different Generative AI systems used to make content. Each technology has pros and cons, so users and businesses must be cautious and know much about it.

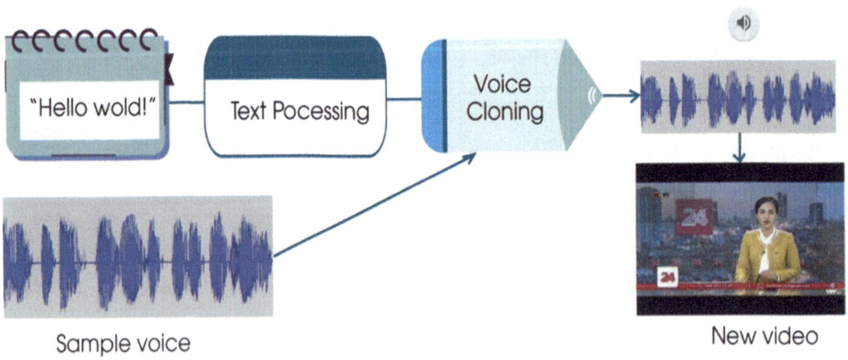

Fig. 5 The steps for making a TTS, voice clone, and video news bulletin

Table 2 A look at generative AI technologies and how they might be used to make content

Type of technology	Main description	Important tools and platforms	Main uses in journalism and making media material	Key challenges/ considerations
Text generation (LLM, NLG)	Based on enormous language models that have been trained on huge datasets, it makes natural, coherent writing from data or prompts	GPT-4, BERT, T5, WriteCream, Typli.ai, Easy-Peasy.AI	Writing short news stories, sports news, financial reports, summaries of texts, blog posts, and scripts	The editor's job, quality, accuracy, the possibility of "hallucination," quick engineering skills, and plagiarism
Chatbots and Interactive storytelling	Makes conversations interactive, helps users, collects information, and conveys tales in an interactive way	Dialogflow, Rasa, Character.ai, ASR-integrated chatbots	Helping reporters gather and summarize information, talking to readers, giving them individualized information, conveying stories in an interactive way, and getting comments	The safety of user data, the accuracy of acquired information, the ability to understand complex situations, and the ethics of interaction
Image and Infographic generation	Using models like GANs or Diffusion, it turns written descriptions or data into pictures, visuals, and infographics	DALL-E 2, Midjourney, stable diffusion, venngage, Piktochart	Making pictures for social media, making infographics that show data in a clear way, making advertising banners, and making pictures for articles	Image quality, the risk of deepfakes, image copyright, the veracity of data in infographics, and comprehensive prompt requirements

(continued)

Table 2 (continued)

Type of technology	Main description	Important tools and platforms	Main uses in journalism and making media material	Key challenges/ considerations
Video generation	It makes films from written scripts, photos, or other material assets on its own, and it usually comes with AI voiceovers and subtitles	Pictory, Synthesia, InVideo, Designs.ai, Veed.io, Heygen, Vidnoz	Making short news bulletins, videos that summarize events, videos that explain things, videos that promote things, avatar videos, and videos for social media	Video quality (how natural the movement and voices are), the risk of deepfake videos, the copyright of the images and audio used, restricted originality, and emotion
TTS and Voice cloning	Text-to-speech (TTS) turns text into natural speech, and voice cloning makes a digital copy of a real person's voice	Google TTS, Amazon Polly, and Play.ht, Verbatik, ElevenLabs	Making audiobooks, podcasts, voiceovers for videos, automated radio news, voice assistants, and audio experiences that are unique to each person	Voice quality, naturalness, expressiveness, copyright of voice, possibility of scams and impersonation using audio deepfake, and complicated ethical and legal considerations

3 Applications in Journalism

The fast growth of AI, especially generative technology, is changing how journalists do their jobs. Artificial intelligence is used in many parts of the news-gathering process, such as keeping an eye on data, summarizing content, suggesting headlines, and finding trends. These things help newsrooms do their jobs better and get this personalized content to the right people.

Like other types of mass communication, journalism has rules that it must follow. However, it emphasizes ethics, objectivity, and social responsibility more. Human journalists do important work, such as investigative journalism, putting things in context, and handling sensitive information. You need to be able to think critically, have experience, and make editorial decisions for these jobs. At this point, AI can only help with journalistic work; it cannot do the vital things people do.

3.1 Improving the Way News is Made

One of the most obvious ways that AI has helped journalism is by making several steps in the news production process more efficient and automated. AI can help with almost every step, from finding potential trending topics and events online early on, to helping with data collection and filtering from many sources, to analyzing information, to writing basic news reports, and even to automating the publishing process for some types of content [10]. Many big news organizations were the first to use AI to make their operations more efficient. For instance, Reuters, one of the biggest news agencies in the world, uses AI techniques mostly to speed up news distribution. "Fact Genie" is a well-known internal tool that can quickly read press releases and source materials and then suggest important alert headlines for the newsroom to work on. AI can make suggestions, but it is up to people to decide whether or not to publish something. This ensures that the information is correct and that the editors are to blame [11]. The Associated Press (AP) has also started an AI project to help newsrooms that are not as big. This project will help them automate routine tasks like processing information from police reports, making quick summaries of live events (like at KSAT-TV), automatically making weather forecasts (like at El Vocero in Puerto Rico), or sorting and classifying emails from readers that could be about topics (like at WFMZ-TV).

AI is especially good at making news from structured data, including sports news (updating match results and standings) and financial news (changes in the stock market and company financial reports). You can teach AI systems to automatically get data from these sources, update the most recent numbers, and quickly and accurately write basic articles that describe the data. Figure 6 shows how AI Automation is used in journalism and the media to automate news production, from writing articles and making videos to creating virtual newscasters and customizing information for each user.

This automation makes everything more efficient and helps newsrooms, especially smaller ones, save costs and shift personnel to other important responsibilities, including in-depth investigations. However, human responsibility in evaluating "news value," providing comprehensive context, and minimizing the amplification of inherent biases in data by AI is an essential function.

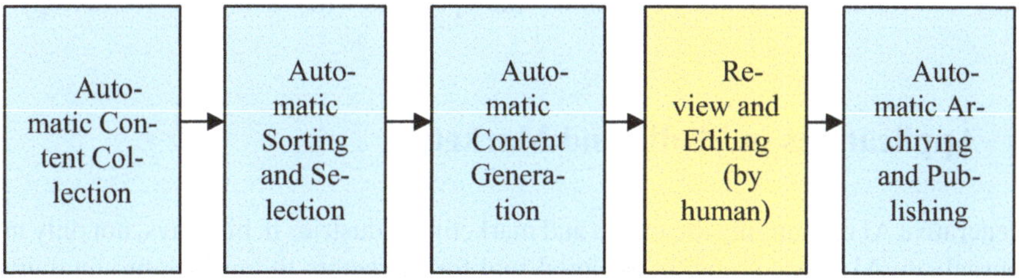

Fig. 6 The process of collecting and making news automatically with AI automation

3.2 Content Summarization and Editing

AI is essential in supporting editors. It can automatically summarize lengthy texts, helping journalists quickly grasp key points. AI systems can also pull out important quotes, confirm facts by comparing them to big databases, and come up with interesting categories, keywords (tags), or headlines for articles that will help with SEO. AI makes computerized summaries of live events at KSAT-TV, for instance, which frees up reporters to do other things. AI techniques can also help you find important phrases and statements in large texts or recorded interviews. This technology saves editors a lot of time, which lets them focus on making journalistic products better [12].

But if you rely on AI to check facts, you could potentially have "automation bias" if the AI model "hallucinates," which means it makes up information that isn't true. So, people are still responsible for making sure that all public information is correct.

3.3 A/B Testing and Personalization

Newsrooms are putting more and more effort into personalizing user experiences and improving content using scientific testing methodologies [13] to get and keep readers in an information landscape that is becoming more and more competitive. AI is a big part of both of these things. AI can assist in making alternative versions of news articles that are better for different categories of readers by looking at how users behave. AI is also an excellent tool for A/B testing, which helps newsrooms figure out which versions of their content (headlines, photos, etc.) work best. This helps them keep refining their content strategies [14].

Howerver, too much personalization can create "filter bubbles," which stop readers from seeing other points of view. But A/B testing might make people act like "click-bait" if it only looks at how to get more clicks. This makes journalism less reliable and less good. Table 3 shows some interesting survey results from 2024 that give a more detailed view of how AI is used in journalism.

Table 3 describes a complicated picture: Newsrooms are quickly and aggressively adopting AI technology as an important tool for the future. However, the public is still wary and wants high standards of ethics and openness when using this technology.

4 Applications in Media and Marketing

Generative AI is changing the media and marketing industries in big ways, not only in journalism. AI is becoming an essential tool for marketers to improve productivity and reach corporate goals. It helps them establish content strategies that work on

Table 3 AI usage rates and types in newsrooms around the world (based on a 2024 survey)

Question for the index/survey	Percentage or result %	Source/note
The percentage of newsrooms throughout the world that use AI technology	73%	Sonni et al. (2024a)
Percentage of leaders of digital news organizations who think that investing in AI is important for journalism to stay alive	78% of the 200 leaders who were asked	Newman (2024), Reuters Institute Digital News Report 2024
How many Americans think that U.S. news media use AI (at least sometimes)	About 67% (two-thirds)	Based on survey data from 2024, the MJC Report (2025)
Trust in newsrooms that use AI	Three times as many people have little or no trust as those who have some or a lot of trust	MJC Report (2025)
People want to use AI chatbots to get the news	Almost half don't care, and 40% might utilize it if the news organization checks the facts	MJC Report (2025)
Readers want to know how AI is applied in journalism	About half of the people feel that disclosure is "very important," even if reporters and editors are still checking the facts. 58% want publishers to have clear rules around ethics before they try out AI	MJC Report (2025)
Newsrooms use AI a lot (examples from AP)	Automated transcription and live event summarizing (KSAT-TV), automated weather bots (El Vocero), email subject sorting (WFMZ-TV), picture generation (Baltimore Times), expert source searching (Haitian Times)	AP Case Studies

multiple platforms, automate advertising and SEO, analyze user behavior, and create quick feedback loops.

4.1 Management of Production Across Many Platforms

AI is a powerful way to deal with this complexity in a world where customers use many different platforms. AI may help make and change content to fit the needs of each platform. For example, it can turn a big blog post into a short video for TikTok, pictures for Instagram, or long pieces for LinkedIn. AI algorithms can also assist you in finding the best times to post to get the most engagement [15]. This helps marketers keep their brands consistent and carry out their plans more efficiently. However, if there isn't a good content strategy in place, this automation might make "information noise" worse. So, people must still be at the center of creating real value.

4.2 Automated SEO and Ads

AI is making big strides in advertising and SEO by automating hard jobs. Generative AI can automatically make several versions of ads for A/B testing or make better meta descriptions for SEO. AI systems like BrightEdge help with keyword analysis and figuring out what users want, which greatly boosts site traffic and sales. AI can also help you come up with "hot" subjects for making content [16]. However, success still depends on the entire plan and how well people understand it; AI is a strong tool, not a "plug-and-play" solution. If you do not handle it correctly, making a lot of SEO content might lower the quality and get you in trouble with search engines.

4.3 A Loop for Behavior Analysis and Feedback

One way AI might be used in marketing is to collect and analyze data on how users behave and provide quick ways for users to give feedback. They may take in many data from different encounters and use it to make predictive behavior models that help marketers understand and predict customer trends. AI may also help create automated reporting systems that let you see campaign results in real time and give you ideas on how to make them better. Collecting personal information raises privacy issues, and there are many ways that it might be misused to trick customers. Relying too much on short-term analytics makes it more likely that marketers will overlook brand-building goals or unique creative components.

Table 4 gives a summary of some important numbers on how AI is used and how people in the marketing sector see it. This is based on recent facts and predictions.

Table 4 clearly illustrates that AI is becoming more important in marketing. Many people are using it and think it is helpful for productivity and creativity. However, there are still problems with data privacy, ethics, and integration.

5 Important Benefits

Using AI to make and share content has many important benefits that change how journalism and media work [17, 18]:

Table 4 Statistics on AI usage and perception in marketing (2025 and forecast)

Question from the index/ survey	Percentage percentage or result %	Source/note
How many marketers use AI tools	73% (expected to go up by 2024)	SEO.com, Semrush (2024)
Marketers who think AI makes people work faster	80%	SEO.com, Semrush (2024)
Marketers who think AI makes people more creative	60%	SEO.com, Semrush (2024)
The best ways to use AI in marketing	Making content (text, photos, video), analyzing data, customizing things, and helping customers	Salesforce (2023)
The hardest things about employing AI in marketing	Data privacy, moral issues, expense, and how well it works with current systems	IBM (2023)
Forecast for the size of the global AI in marketing market	Expected to grow at a rate of 32.5% each year (2023–2028) and reach USD 107.5 billion by 2028	Market sand Markets (2023)

Cost savings: AI automates repetitive operations that take time, such as collecting data, writing basic articles, and making routine material. This cuts down on personnel and operating costs for newsrooms and media companies.

Time savings: AI makes it possible to quickly create content, especially for data-driven news (such as sports scores and financial reports) or to make numerous ad versions quickly. This reduces production time and lets businesses quickly respond to market needs.

Ability to scale content production: With AI, businesses can make more content that works better than traditional techniques. AI can constantly search for and organize huge volumes of information across many platforms without hiring more people.

Content optimization for each platform: AI looks at data to determine how people behave on different platforms. Then, it helps you customize your content's formats, styles, and posting times to get the most views and engagement on each platform.

Better user experience through personalization: AI makes it possible to give highly personalized content that is most relevant to users' interests and preferences, which leads to more engagement and retention.

These advantages are not just preventative; they are also essential from a strategic point of view since they allow companies to make important changes to their business models and find new sources of income. However, to truly comprehend these benefits, one needs to invest in training for technology, data, and people.

6 Limitations and Risks

There are many pros and downsides to using AI in content development and dissemination, and these need to be carefully thought about [19–21]:

Delusion and the risk of false information: Generative AI models can sometimes create "hallucinations," which are information that is wrong or entirely made up but looks real. That may easily lead to false information and a loss of trust in the media, which is a big problem.

Quality and depth of content: AI can make much content, but the quality, creativity, depth, and capacity to transmit emotion or critical thinking are frequently not as good as content made by people, especially for creative narrative or complicated investigative journalism.

Intellectual property and copyright: Using copyrighted data to train AI and make new content presents a lot of complicated legal and moral considerations about intellectual property rights. It is still hard to determine who owns and can use AI-generated content.

Ethics in AI—Bias, Privacy, and Responsible AI Development: AI models are trained on historical and present data, which might show and potentially make worse societal biases (such as race, gender, and culture). AI-generated material can continue to prejudice, promote preconceptions, and create "filter bubbles" or "echo chambers" if it is not carefully managed. These limit consumers' exposure to different points of view.

Job loss and changes in the workforce: People are worried that AI will take over more jobs in journalism and the media. This means that the workers there need to change and learn new things.

Too much dependence and insufficient critical thinking: People who rely too much on AI might be unable to think critically, look at information carefully, or check facts. This makes it hard to find and fix AI's errors or biases.

Data privacy and security: Using user data for personalization and behavioral analysis is a significant threat to privacy and cybersecurity.

Table 5 shows that AI has two sides when creating and sharing material. This shows that we need to prepare ahead and have strong ethical guidelines to get the most out of AI while avoiding any problems.

7 Automated Distribution and User Experience

AI can do more than only make content; it can also improve how content is shared and how users interact with it [12, 22].

Table 5 Comparison of benefits and risks of AI in content creation and distribution

Aspect	Benefits	Risks	Some mitigation measures/Exploiting benefits
Efficiency and Scale	Saves cost and time Quick content creation The ability to scale up production	The risk of "information noise" from making too many copies Too much reliance on technology might stifle human ingenuity	Put more emphasis on quality than quantity Editorial decisions are made with human scrutiny Use AI to do boring duties so people can do creative work
Content quality and Accuracy	Optimized for each platform Supports basic content generation	"Hallucination" (wrong knowledge) Content that isn't deep or emotional Training data that makes bias worse	A strict assessment and fact-checking by people AI gets a lot of different and high-quality training data Make Explainable AI (XAI) to be clear
User experience and Interaction	Enhanced personalization Interactive storytelling More people are using it	"Echo chambers" and "filter bubbles" Worries about privacy and how data is used The chance of "clickbait" optimization	Be open with users about how AI is used Put ethical AI development (Ethical AI) at the top of your list Get content from a variety of places and promote critical thinking
Legal and Ethical aspects	Made operations more efficient	Copyright violation and intellectual property Fake news, impersonation Loss of jobs	Clear rules for AI-generated content in the law Strong protection for original works under copyright law Teaching workers new skills and retraining them

7.1 Systems that Suggest Personalized Content

AI-assisted recommendation systems are becoming increasingly important since they can tailor how material is shared. AI can make wise choices when proposing articles, videos, or podcasts by looking at users' past interactions and engagement, such as their viewing history and search terms. This level of customization gets users to spend the most time on the platform and be the most engaged. For instance, Netflix and

Spotify use similar technologies to suggest new content to users. A news app might also use a recommendation system to get each user to connect with the products, giving them individualized news feeds and making it more likely that they would keep interacting with the information.

7.2 Scheduling and Sending Content Automatically

AI tools can help you figure out when and where to post content on different channels. For instance, an article could be put on a website, shared on social media, sent out in a newsletter, and so on. AI may help determine when an audience is most engaged, increase reach, and suggest the best timing and number of times to share certain content. It might look at audience demographics, time of day distributions, trends, past content engagement statistics, and more to develop a better publication schedule.

7.3 Analytics in Real Time and Improving Performance

AI systems keep track of performance measures like completion rates, engagements, and conversions to give you real-time statistics on how well your content is doing. We use this information to give users insights and make suggestions on how to change their content and distribution tactics in real time. This "feedback loop" lets media companies and advertisers adjust their plans rapidly and improve campaigns or content so they work better.

7.4 AI in Making Audiobooks and News Broadcasts

AI is transforming how audio material is made. As we discussed, TTS technologies let people listen to articles instead of reading them by turning them into audiobooks or news bulletins. Voice cloning lets news organizations automatically turn text into audio using voices that people already know for automated broadcasts, which makes listeners more familiar with the news. Play.ht, Verbatik, and ElevenLabs are three well-known tools that use this idea.

7.5 AI for Managing Communication Campaigns

AI can do more than only make content; it can also help with campaign management. AI-powered platforms can keep an eye on brand mentions across the web, figure out how people feel about a company, find influencers that impact a brand,

and answer customer questions automatically in certain circumstances. This can make customer service more efficient, improve the business's reputation, and make it easier to manage communications, such as crisis communication or significant public relations initiatives.

8 Trends and Research Directions for the Future

To fully use AI's promise and reduce its hazards, several significant trends and research topics are being worked on [23, 24]:

Explainable AI (XAI): Researchers are making AI models easier to grasp and see through so that people can see how AI makes decisions and creates information. That is very important for building trust, especially in journalism, where being responsible is the most important thing.

Responsible AI development and ethical AI: More and more people are saying that we need to make AI systems that follow moral rules like fairness, privacy, and not discriminating against people. This means thinking about moral issues at every stage of making and using AI.

AI-Human collaboration (Human-in-the-Loop): The future of AI in media will probably feature a strong collaborative approach in which AI does repetitive, data-heavy jobs while people do things like creative work, critical thinking, ethical oversight, and making strategic decisions. Research will investigate the most effective human-AI interfaces and workflows.

Personalized content at scale: Advanced AI will allow for even more personalized content, giving each user exactly what they want while ensuring that there is a variety of information to avoid "filter bubbles".

Multimodal generative AI: Future AI models will be better at simultaneously making content in more than one way. For example, they could make a video from a text prompt that contains spoken dialogue, graphic scenes, and background music. This will make multimedia production even easier.

AI in investigative journalism: Research will look into how AI might help with difficult investigative jobs, such as looking through huge datasets to find patterns, anomalies, or links that human reporters might overlook, all while following rigorous ethical rules.

Regulatory Governance for AI: As AI improves, we must quickly implement the correct laws and rules to safeguard creators and consumers of AI-generated content, copyright, deepfakes, and data privacy.

Evolving skillsets for communication professionals: Communication workers need to learn new skills beyond what they already know as AI continues to change the media landscape. Some important abilities are making good prompts, judging AI-generated content fairly, and knowing about digital ethics, privacy, and intellectual property challenges. These skills enable people and AI to work together responsibly and helpfully in the post-digital age.

9 Conclusion

Chapter 4 looks at how AI changes how news and media companies make and share content. In doing so, it shows a wide view of the profound changes that modern technology has caused.

The chapter mainly discusses how Generative AI could change how things are done. Technologies like LLMs that help with writing, chatbots that make interactive stories, and multimedia generation tools that make images, videos, and audio are all signs of a new era in content creation. Their use allows media companies to improve their efficiency, save time and money, and increase their production capacity while also being able to easily change content for different distribution platforms.

The chapter then looks at real-world uses of AI in the news ecosystem. The technology is presently being used to make production workflows more efficient, help journalists write, shorten, and edit text, and make A/B test reader experiences more personalized to boost engagement. AI's presence in the larger media and marketing world makes it easier to coordinate production across platforms, automates advertising and SEO operations, and gives us detailed information on user behavior with feedback loops that happen almost instantly.

The Chapter also has a long part that gives a balanced view by critically looking at the problems and hazards of using AI. Some of the main problems discussed were the risk of AI "hallucinations" spreading false information, the possibility of material quality and depth going down, and complicated legal battles over copyright and intellectual property. It also carefully looks at ethical problems such as biases built into data, societal effects like job loss, the loss of critical thinking skills due to too much reliance on technology, and significant data security and privacy issues. The chapter asserts that effective AI utilization necessitates firms to implement a proactive risk management approach, continually emphasizing human monitoring and adhering to the idea of transparency.

In conclusion, it is essential to develop a complete strategy framework if you want to use AI's full potential in journalism and media in a way that lasts. This plan should be based on two ideas: encouraging strong technological innovation while upholding the fundamental values of accuracy, trustworthiness, and professional ethics. This method will not work unless everyone involved is committed to being accountable. This means that all important stakeholders need to work together. The people who make technology have a basic duty to make AI systems clear and understandable and actively work to reduce biases built into them. Media companies, conversely, need to set clear rules for using AI responsibly, strategically train their employees, and ensure that human oversight is always the most important thing. At the same time, governments need to quickly build flexible legal and regulatory frameworks to protect the rights of stakeholders and oversee AI applications effectively. Ultimately, the user community needs the right critical literacy and abilities to be active and discriminating information consumers in the digital age.

AI is making it possible to automate the development and dissemination of material. That trend cannot be stopped; it has much potential and many problems. For the journalism and media industries to grow in a way that is good for society in the future, they will need to understand the technology, see opportunities and risks clearly, and, most importantly, put people at the center of this change.

Declaration of Generative AI and AI-Assisted Technologies in the Chapter Preparation Process During the preparation of this work the author(s) used Gemini/Grammarly in order to support language editing, improve clarity and coherence of the text, and assist in restructuring sentences while preserving the original scientific meaning. After using this tool/service, the author(s) reviewed and edited the content as needed and take(s) full responsibility for the content of the published article.

References

1. Sengar, S.S., et al.: Generative artificial intelligence: a systematic review and applications. Multimed. Tools Appl. **84**(21), 23661–23700 (2025)
2. Becker, J., et al.: Text generation: a systematic literature review of tasks, evaluation, and challenges. arXiv preprint arXiv:2405.15604 (2024)
3. Devlin, J., et al.: BERT: pre-training of deep bidirectional transformers for language understanding. In: Proceedings of the 2019 Conference of the North American Chapter of the Association for Computational Linguistics: Human Language Technologies, vol. 1, pp. 4171–4186 (2019)
4. Veglis, A., Maniou, T.A.: Chatbots on the rise: a new narrative in journalism. Stud. Media Commun. **7**(1), 1–6 (2019)
5. Zhang, C., Zhang, C., Zhang, M., Kweon, I.S.: Text-to-image diffusion models in generative AI: a survey. arXiv preprint arXiv:2303.07909 (2023)
6. Orak, C., Turan, Z.: Using artificial intelligence in digital video production: a systematic review study. J. Educ. Technol. Online Learn. **7**(3), 286–307 (2024)
7. Yu, T., Yang, W., Xu, J., Pan, Y.: Barriers to industry adoption of AI video generation tools: a study based on the perspectives of video production professionals in China. Appl. Sci. **14**(13), 5770 (2024)
8. Tan, X., Qin, T., Soong, F., Liu, T.Y.: A survey on neural speech synthesis. arXiv preprint arXiv: 2106.15561 (2021)
9. Genelza, G.G.: A systematic literature review on AI voice cloning generator: a game-changer or a threat? J. Emerg. Technol. **4**(2), 54–61 (2024)
10. Liu, X., Nourbakhsh, A., Li, Q., Shah, S., Martin, R., Duprey, J.: Reuters tracer: toward automated news production using large-scale social media data. In: 2017 IEEE International Conference on Big Data (Big Data), pp. 1483–1493. IEEE (2017)
11. Sonni, A.F., Hafied, H., Irwanto, I., Latuheru, R.: Digital newsroom transformation: a systematic review of the impact of artificial intelligence on journalistic practices, news narratives, and ethical challenges. Journalism Media **5**(4), 1554–1570 (2024)
12. Islam, M.A., Fakir, S.I., Masud, S.B., Hossen, M.D., Islam, M.T., Siddiky, M.R.: Artificial intelligence in digital marketing automation: enhancing personalization, predictive analytics, and ethical integration. Edelweiss Appl. Sci. Technol. **8**(6), 6498–6516 (2024)

13. Yoneda, T., et al.: Algorithms and system architecture for immediate personalized news recommendations. In: IEEE/WIC/ACM International Conference on Web Intelligence, pp. 124–131 (2019)
14. Quin, F., Weyns, D., Galster, M., Silva, C.C.: A/B testing: a systematic literature review. J. Syst. Softw. **211**, 112011 (2024)
15. Mokale, M.: Customizable software tooling for managing cross-platform media content delivery. Int. J. Multidiscip. Res. **7**(1) (2025)
16. Sharma, A., Dhiman, M.P.: The Impact of AI-Powered Search on SEO: The Emergence of Answer Engine Optimization. https://doi.org/10.13140/RG.2.2.20046.37446 (2025)
17. Holzner, N., Maier, S., Feuerriegel, S.: Generative AI and creativity: a systematic literature review and meta-analysis. arXiv preprint arXiv:2505.17241 (2025)
18. Katsamakas, E., Pavlov, O.: AI and business model innovation: leveraging the AI feedback loop. J. Bus. Models (JOBM) **8**(2), 22–30 (2020)
19. Bateman, J.: Deepfakes and Synthetic Media in the Financial System: Assessing Threat Scenarios. Carnegie Endowment for International Peace (2022)
20. Dang, L.M., Hassan, S.I., Im, S., Lee, J., Lee, S., Moon, H.: Deep learning based computer generated face identification using convolutional neural network. Appl. Sci. **8**(12), 2610 (2018)
21. Gupta, M., Akiri, C., Aryal, K., Parker, E., Praharaj, L.: From chatgpt to threatgpt: impact of generative AI in cybersecurity and privacy. IEEE Access **11**, 80218–80245 (2023)
22. Lim, G., Perrault, S.T.: December. XAI in automated fact-checking? The benefits are modest and there's no one-explanation-fits-all. In: Proceedings of the 35th Australian Computer-Human Interaction Conference (2023), pp. 624–638
23. Ozurumba, C.J., Fagbohun, O., Akintola, A.S.: Ethically Integrating AI Into Journalism: Balancing Innovation, Integrity, And Public Trust. Integrity, And Public Trust (2024)
24. Pawar, D.: Immersive storytelling: artificial intelligence, virtual reality and the future of audience engagement. In: Digital Pulse: Shaping the Future of Media and Communication, pp. 56–66 (2025)

Emotional Intelligence in Digital Human-AI Interaction: A Meta-Analysis of Empathy, Trust, and Engagement

B. Pathinettampadiyan, H. Samuel Thavaraj, and V. Seema

Abstract AI systems increasingly interact with users on an emotional level. The simulation of emotional intelligence has become a fundamental aspect of human–AI design. This meta-analysis synthesizes findings from 32 empirical studies (N = 4720) investigating the relationship between elements of emotionally intelligent AI and three major user outcomes: empathy, trust, and engagement. Across various platforms, from chatbots to social robots to voice assistants, emotionally expressive features such as empathic tone, backchanneling, and affective design were strongly related to increased user responses. The analysis revealed a moderate, consistent effect size ($r = 0.31$), suggesting that emotional intelligence consistently impacts socio-affective experiences of human–AI interaction. Implicit emotional cues were strongest, often outperforming explicit self-report measures. While benefits are evident, ethical problems exist, such as emotional manipulation, anthropomorphism, and cultural conflict. Implications of this work contribute to theory and practical advice toward creating ethically responsible, emotionally intelligent artificial intelligence systems.

Keywords Emotional intelligence · Human-AI interaction · Meta-analysis

B. Pathinettampadiyan
Full TIime Research Scholar, Department of Rural Industries and Management, Gandhigram Rural Institute - Deemed to be University, Gandhigram, Tamil Nadu, India
e-mail: rs24913408@ruraluniv.ac.in

H. Samuel Thavaraj
Associate Professor, Department of Rural Industries and Management, Gandhigram Rural Institute - Deemed to be University, Dindigul, India
e-mail: h.samuelthavaraj@ruraluniv.ac.in

V. Seema (✉)
Assitant Professor, Department of Rural Industries and Management, Gandhigram Rural Institute - Deemed to be University, Dindigul, India
e-mail: drseemagri@gmail.com

P. Chien Thang and A. Prasanth (eds.), *AI-Assisted Journalism and Media: Opportunities and Challenges*, Studies in Computational Intelligence 1277,
https://doi.org/10.1007/978-3-032-19648-4_5

1 Introduction

A breakthrough in affective computing has facilitated the advancement, a multidisciplinary area of study that is a convergence of computer science and emotional science for creating systems capable of recognizing, understanding, and simulating emotions. Emotional Intelligence, once viewed as a solely human cognitive-emotional capacity [31]. The evolution has been made possible by advancements in affective computing, a field of study that combines emotional science, analyse, and simulate emotions [69]. Through this, the users can sense an emotional connection, viewing machines as reliable or empathetic. In some cases, users can invest human-like attributes to these agents, thus establishing a difference between genuine emotional relationships and artificial implementations [19, 24].

These developments imply increased interaction and access in human-computer interaction, but also bring serious psychological and ethical issues. Can a machine understand human emotion, and can an artificial intelligence system simulate emotion to build trust and satisfaction, but also the potential for manipulating affect, where systems deploy emotional responses without true emotional capabilities [52]. This is at once a philosophical and practical problem of emotional truth because users are impacted by machines that simulate care and empathy without intentions, consciousness, or ethical responsibility [16].

Emotional Intelligence is a twofold part in the domain of human-AI interface, referring to the emotional capability of users as they experience and respond to AI behaviors, and also to the emotional expressiveness deliberately embedded in AI systems. In either role, EI is instrumental in shaping user experiences and perceptions. Systems with emotional intelligence are often seen as more competent, trustworthy, and sociable, especially in scenarios that demand emotional sensitivity, such as mental well-being services, care for the elderly, or learning environments [7, 25]. Empathy, a core feature of EI, goes some way toward making users feel at ease and understood, while engagement, framed as emotional concern and attentional investment likely to rise when users feel that the AI understands their emotions or responds to their emotional cues appropriately [42]. The extensive body of works on human-to-human interaction expressions that high emotional intelligence is a core factor in improving communication, leadership, and interpersonal relationships [38, 67]. Endowing machines with similar emotional abilities, such as affective speech, gesture mimicry, and voice modulation, designers seek to emulate the emotional potency of human interaction. As artificial as these dense emotional cues are, they can effectively elicit a high degree of social presence, making users perceive that their interactions occur with sentient partners [4]. This technique has delivered successful results in digital learning environments and customer service, where affective bots boost motivation, reduce frustration, and improve user satisfaction [29].

Results need to be interpreted with caution. Researchers warn that humans, especially those in emotional distress, are prone to anthropomorphising or attributing emotional or moral agency to expressive AI that does not possess such capabilities [83]. Furthermore, there is no compromise on the definitions and measurements of

AI Emotional Intelligence (EI). Some see it as just the simulation of emotion, while others call for actual emotion detection and that the system respond with situationally appropriate emotion [34]. Results from studies on user-related outcomes, such as trust, empathy, and engagement, are so dissimilar that the outcomes are fragmented and contradictory [5]. Social and cultural variables such as the user's culture, age, or the appearance of the AI (human-like robot vs. text-based bot) also confound the interpretation of results [89].

Despite the high importance of affective dynamics in AI research lacks a quantitative and integrative synthesis of the empirical findings obtained so far and most of the paradigm seems to be confined to waves of research within specific academic schools of thought preventing the development of a synoptic view of the psychological and practical implications of machines with emotional intelligence. There is an urgent need for the synthesis of such findings, the determination of the robustness, consistency and possible moderators of effects of Emotional Intelligence in AI systems. Emotional Intelligence in Human-AI Interaction, it is necessary to address the increasing complexity and social integration of AI systems mimicking affective behaviour [51]. Without consciousness and subjective experience in machines, designers seek to emulate the behaviour al expression of such capabilities to allow for emotionally congruent interaction with users [71].

One of the most challenging aspects of this task is translating intrinsically human emotional processes into plausible and ethical computational symbols. For instance, affective Natural Language Processing technologies attempt to measure user sentiment modify AI responses in ways appropriate to emotional suitability [92]. These technologies rely heavily upon emotionally labelled datasets with valence, intensity, and contextual salience, such as sentiment lexicons and emotion-labelled corpora, that allow the AI to mimic responses appropriate to human emotional expectations [84]. Despite practical use, these simulations are superficial copies, without genuine emotional depth.

The variance of cultural and subjective differences in expressing and understanding emotions is one major challenge facing the universal implementation of emotional intelligence in creating artificial intelligence. What is empathetic or emotionally apt in one culture can be perceived as intrusive or insincere by another culture [53]. The global implementation of emotionally intelligent AI systems should take into account socio-cultural nuances to avoid emotional conflict, which can cause loss of user trust or miscommunication [74].user expectations of emotionally intelligent AI are also shaped by the system design and human natural leaning [57]. Empirical data show that people interact with emotionally expressive agents in a human-like manner, attributing them with such human-like attributes as intentions, awareness, and empathy, despite their awareness of the artificiality of the agent [88].

2 Theoretical Background

Emotional Intelligence developed from a first proposal of how people think with feelings to a multifaceted construct at the connection of psychology, neuroscience, communication. The Early Ability Model situated EI as an information processing hierarchy, perceiving, using, understanding, and regulating emotion, supplementing rather than opposing cognition [31]. Empirical studies thereafter found that each branch has discriminant and incremental predictive validity for social adaptation, work performance, and well-being. Ability-based EI measures, such as the MSCEIT, created controversy about whether emotional difficulties have objective solutions [94]. The Mixed Model extended the frame by integrating cognitive, affective, and motivational skills with empathy and social influence [31]. Popularization of this led to organizational take-up but also diffused distinctions between EI, personality, and acquired skills. Authors warned against conceptual diversity and measurement inflation [93]. In Trait EI as a set of habitual self-perceptions within personality hierarchies, measured by self and predictive of resilience, coping, and relationship quality [68]. These streams of work suggest that EI is not one but a family of constructs whose relevance depends on whether the research question is concerned with maximal performance, typical behaviour, or developmental potential [55].

The background attributes of the human user that shape expectations of technology and as system design elements constructed into the system to elicit social emotional responses. The latter draws on the intellectual heritage of affective computing [69]. They intended to prepare computers with the dimensions to sense, where relevant, generate affective signals. Drawing on classical emotion appraisal theory [81]. Affective computing systems employ multimodal sensors like facial action coding, prosodic vocal analysis, galvanic skin response, eye gaze, and even keystroke dynamics to predict user affect with increasing accuracy [9]. Algorithms then interpret these predictions as adaptive responses: rephrasing a chatbot, adjusting the difficulty of a tutoring system, or adjusting a robot's proxemic behaviour. Although these responses give the impression of emotional understanding, they are phenomenologically indifferent and based on pattern matching over labelled data.

Empathy theory accounts for this tension. The distinction between cognitive perspective taking and affective sharing has guided psychological measurement, like the Interpersonal Reactivity Index, and computational models attempting to infer mental and affective states [17]. Synthetic empathy, implemented in social robots or conversational agents, typically combines multimodal cues to select empathic response templates or behaviours, facial expression matching, or supportive speech [32]. User studies consistently show that such cues enhance perceived warmth and relational satisfaction [42]. Neurocognitive investigations complicate cognitive science, mirror system activation and oxytocin release are elicited by robot companions, blurring neural boundaries between human and artificial social stimuli [89]. Ethical standards now demand open disclosure of artificial identity, regulation of persuasive design, and safeguarding against emotional manipulation. Trust, known since long ago as a social cooperation lubricant, works in HAI through a calibration frame that the users need to calibrate trust to the competence and intentions of the system [40]. Emotional expression acts as a heuristic for that calibration. Individuals routinely use social rules, politeness, reciprocity, and even stereotyping to interact with media that present humanlike cues [57]. When an AI apologizes for mistakes with an apologetic tone or adjusts feedback to user frustration, it triggers social scripts of goodwill and integrity, thereby enhancing trust [6]. Culturally incompatible emotional expressions, on the other hand, lower credibility and eye contact standards or display rule expectations vary in collectivist vs. individualist cultures, influencing robot acceptance [58]. Embodiment introduces yet another element: embodied robots evoke more substantial social presence, yet also create expectations of rich emotional display over disembodied voice assistants. Ultimately, trust in AI is co-constructed by technical competence, reasoning transparency, and the interface's fine-tuned emotional choreography.

In emotional design, principles synthesise these discoveries into a three-part model: visceral response to sensory aesthetics, pleasure through usability, and reflective meaning connected to identity and values. In practice, AI designers use colour, animation smoothness, voice tone, and narrative style to elicit positive feelings on all levels. Gamified mental health apps, like Warbot, use empathic language and achievement badges to render cognitive behavioural drills emotionally compelling [22]. Social robot companions, such as PARO the seal, leverage fixed cuteness and calming behaviours to alleviate anxiety in dementia-affected patients. The pursuit of engagement can become a manipulative, affective nudging when financial incentives place engagement metrics above user well-being [96]. Therefore, ethical emotional design requires ongoing user-centred testing and ethical impact assessments.

A blend of EI theory and computational use also involves consideration of measurement validity. Multimethod convergence across ability tests, self-report, peer ratings, and performance outcomes in human research assists triangulation of EI's latent structure [38]. In HAI, corresponding triangulation is achieved by integrating system logged behavioural response, physiological markers, and subjective user ratings to infer perceived empathy or engagement [34]. Operational definition heterogeneity, where trust is operationalized as single-item likability in one study and a composite of reliability and ethical perception in another study, makes meta-analytic synthesis impossible. Standardization efforts, like the Emotion Modelling and Representation Language and the Empathic Cue Set of Gunes, attempt to standardize annotation schemes and outcome measures to enable cross-study comparison. Developmentally, EI in machines has limited data quality and contextual generalization. Western facial expression databases trained deep learning models to misclassify emotions in non-Western faces, perpetuating bias [39].

Ontological limitations notwithstanding, synthetic emotion capability has provided practical utility. In education, affect-sensitive tutoring systems that sense boredom and confusion can adjust the pace, achieving greater learning gain. In customer care, empathy template scripted chatbots lower escalation rates and increase satisfaction scores [30]. In human-robot collaboration, subtle nonverbal signals such as torque-based haptic feedback signal intention and enhance joint action fluency.

These results are consistent with social cognitive theories that emotions are feedback on coordination and decision making [81]. Affective affordances need to be balanced against potential harms. Emotional monitoring using camera streams or keystroke behaviour to infer worker emotions has the potential to invade autonomy and stifle self-expression [16]. Governance systems, like the OECD AI Principles, emphasize human values and resilience, and thus require that developers not only test the technical correctness and the socio-emotional impact. Emotional Intelligence offers a broad theoretical structure in its inclusiveness in understanding and designing the emotional aspect of human-AI interaction. A solid evidence base backed by systematic, standardized measurement and interdisciplinarity can guide designers, researchers, and policymakers in developing emotionally intelligent technologies that respect human dignity and augment, not exploit, our capacity to feel and connect.

2.1 Empathy in Emotional Intelligence and Human–AI Interaction

This research investigates the role of digitally mediated empathy through AI companions (such as Replica) in helping Chinese women to cope with the pandemic. The research, by applying empathy frameworks and the resilience communication theory, shows how empathic encounters with AI prompted the participants to treat AI as a coping resource. Features included a friend, an mood management program, a reactive diary, a digital pet, and a venting mechanism. Each category had differences in cognitive and affective empathy. The results generalize empathy theory from human-to-human to human-AI communication [36]. Artificial Intelligence is revolutionizing interactions between businesses and consumers. However, most AI marketing representatives lack emotional warmth and human sympathy. To circumvent this problem, the article highlights the need for artificial empathy as a key design component of forthcoming AI marketing systems. By combining knowledge of learning from multiple academic fields, the authors suggest a structured approach to incorporating artificial empathy into AI communication. They present an overview of the most significant aspects of artificial empathy and suggest concrete steps for implementation. Empirical research demonstrates that artificial empathy can enhance customer experience and create business value through emotional bridging between humans and machines [45]. Intelligent AI machines are being embedded in different industries at a growing rate, providing speed and cost-efficacy over human labour. Findings indicate that humanlike appearances do not result in trust or acceptance alone. Instead, empathy and interaction quality are crucial mediating variables. Empathetic and interactive AI machines would likely be accepted by consumers [66]. Artificial Intelligence is disrupting service industries so quickly it is difficult to keep up; although efficiencies of operations are offered, customers still have issues with AI, such as lack of trust. Research has shown anthropomorphism and conversations are not enough to establish trust unless they do improve the quality of communication. Empathy response is the most powerful driver of customer trust among service scenarios. The study finds AI's capacity to solidify firm-customer relations, especially during tough periods like pandemics [13]. As conversational interfaces grow more human-like, users treat them more like friends than tools. This study explores how users' mental frames generated by different priming statements about the purpose of the AI affect their interactions. Users were told that the AI was driven by a caring, manipulative, or neutral motivation, and then interacted with the same system. Users primed with the caring intention rated the AI as more reliable, assumed, and helpful. The effect was significantly greater when a more cultured AI model was used, signifying that user perceptions are highly sensitive to framing the AI. A short-term feedback loop was found, where users' mental frames were generated and reinforced through use [65]. Artificial intelligence-based conversational chatbots are widely applied across sectors to engage users with natural language responses. However, most chatbots currently applied on websites or virtual assistants like Alexa and Siri cannot reply with empathy, thus an immature affective responsiveness. While empathy might not be essential for transactional or repetitive responses, it is becoming increasingly essential for chatbots to engage with people in mental health environments where empathetic comprehension and response to users' emotional states are essential [20]. This study examines the integration of artificial emotional intelligence into a social robot named Ryan, designed to assist elderly individuals in coping with grief associated with depression, assesses its effectiveness in user interaction. The empathic Ryan used emotion recognition and expression systems, interpreting facial expressions and speech sentiment to tailor its responses through an affective dialogue manager. In contrast, the non-empathic version relied on scripted dialogues without emotional awareness. Ten older adults in a senior care facility interacted with both versions [1].

2.2 Trust in Emotional Intelligence and Human–AI Interaction

This assessment of human trust in artificial intelligence, in the method of robots, virtual interfaces, or embedded systems, influences this trust. The degree of intelligence of AI affects the formation of trust. Cognitive trust count on AI's reality, transparency, consistency, and proximity, whereas emotional trust is linked to anthropomorphic characteristics. The review also criticizes existing literature for its excessive use of narrow, short-term research [28]. Unlike most cognitive strategy research, this research tested an affective pathway to trust. These results validate that attachment security priming can effectively improve affective trust in AI systems [27]. This research tested trust in robots and AI in first-time interactions through an online trust game. Players trusted a robot "jdrx894" more than a human player such as Michael. Trust was affected by technical education, robot acquaintance, and self-efficacy. Openness to experience increased trust, while conscientiousness decreased it. Results indicate that trust in robots and AI varies based on context, personality, and acquaintance with technology [60]. This research inspects the role of emotional components in the acceptance of AI virtual assistants. Functionality and social emotions were strongly related to trust. Trust, in turn, mediated their effect on user acceptance. Perceived humanity was discovered to have a reversed U-shaped relationship with trust. The research identifies trust to be an important emotional route to AI acceptance and design strategy [95]. The impact of AI perceived humanness on consumer trust in services. Qualitative consumer data were collected through focus group interviews. Anthropomorphism and AI intelligence were the most salient features of humanness under investigation. Results indicate that these characteristics influence trust in a context and realism-dependent manner. Intelligence is desirable, but it induces unease when AI is too human-like. This confirms and develops the "uncanny valley" theory of AI service interactions. The research confirms that a balance needs to be struck to facilitate trust in AI utilization [86]. In two experiments, participants evaluated scenarios in which ATs complied or transgressed ethical standards. Findings showed that transgressions decreased perceived ethicality and trust, but their effect on ethicality was transgression-type dependent. To the same degree as the violation, participants still trusted the AT's task performance, but moral trust was strongly diminished, and the repair attempts were unsuccessful to a large degree [85]. An inclusive conceptual model is proposed for a better empathetic of user trust. The model lays the basis for future empirical tests across different artificial intelligence applications. AI service providers need to understand and establish user trust. Outcomes highlight that establishing trust entails cognitive, emotional, and behavioral aspects. It discusses the key elements, determinants, and consequences of trust in AI from various viewpoints [90]. This scoping review considers conceptualization, representation, and measurement of trust in human–AI interaction across different systems like robots, virtual agents, autonomous cars, and decision support systems. The results show that while trust has been predictable as a vital and multidimensional element of human–AI interaction, there are considerable differences in research approaches used to investigate it. Most of the literature is grounded on a weak theoretical foundation, frequently uses poorly validated trust models, or lacks clear methodological designs [87]. The voice-controlled artificial intelligence personal assistants are increasingly ubiquitous in everyday life, and knowing how trust is established in such technology becomes more crucial. The research indicates a multifaceted relationship between privacy and trust building, with differences in user reaction to the AI assistant's activity and to perceived trustworthiness of the brand in question [70].

2.3 *Engagement in Emotional Intelligence and Human–AI Interaction*

Recognition of the subtleties of emotions is equally important for effective human-robot interaction, especially for care robots. As robots become more combined into our daily lives, it becomes essential that robots are able to sense and respond to emotional human cues. This paper introduces a novel emotion perception system enabling robots to detect emotional states based on facial and auditory signals. Based on recent deep neural network vision and sound processing, this system fine-tunes its responses based on the emotional feedback it receives [21]. Engagement is the interactive and cooperative human-computational relationships that are increasingly becoming a central preoccupation in industrial practice and research. Human-Engaged AI offers an inspiring paradigm for augmenting human potential in cooperation with artificial intelligence systems through integrative collaboration. We view the nature of engagement as a cooperative process of learning between humans and machines [46]. This study presents both qualitative and quantitative results concerning user interaction and emotional relationships established with chatbots. To investigate the complexities of human-chatbot interaction psychological assessments and semi-structured interviews were conducted Results indicated that the individual psychological history of the users heavily influenced the amount of interaction and that users' understanding of emotional behaviours and signals in non-visual interfaces was elucidated and contributes to the existing knowledge small with regard to affective human-artificial intelligence relations and informs on interface design for the future [72]. With the increased usage of chatbots in customer service, it is important to be aware of variables influencing user experience. This study investigates the impact of user experience determinants including anthropomorphism communication style and service scripts on customer experience Employing a qualitative multi-method approach comprising reflection interviews and focus groups the study determines main findings of chatbot performance User satisfaction was significantly affected by the appearance of human-likeness and the tone of communication perceived usefulness and trust were principal predictors of engagement and privacy concerns and negative affect were probable barriers to interaction [2]. AI tools Furnish supportive guidance in writing particularly for target-driven learners with clear learning aims. Feedback literacy in writing is essential for the efficacious application of AI in higher education. This study investigated the impact of AI on L2 writing Motivation and achievement via mixed-methods methodology [73].

3 Methodology

This meta-analysis adhered to PRISMA 2020 guidelines to ensure methodological transparency and replicability at each step of study selection, data extraction, and combination [62]. A systematic literature search was conducted in three prominent databases: Springer, Taylor & Francis, and Google Scholar, to find peer-reviewed empirical studies published between 2000 and 2025. The search strategy involved combining keywords for emotional intelligence, contextual outcomes related to the topic like empathy, trust, engagement, and related technological domains, where necessary, AI systems, chatbots, robots, and virtual agents. Operators were utilized to obtain maximum precision and reach of the outcome. Filters were applied to limit the outcome to English-language quantitative studies on human participants.

The initial database search yielded 1672 records, which were systematically documented in a master Excel spreadsheet with metadata such as author name, publication year, journal, and source database. After removing duplications 388, 1284 unique records were screened at the title and abstract level. Articles were eligible if they measured emotional intelligence using a validated scale or coding scheme and had relevant populations such as entrepreneurs, business owners, or AI system users who used a quantitative method with statistical findings such as correlation or regression coefficients, and reported findings aligned with the objectives of this meta-analysis. Studies were excluded if they were theoretical, qualitative-only, or failed to give extractable statistical information. The full-text screening of 82 eligible articles, 32 studies were retained for meta-analysis.

For all studies in this review, effect sizes were directly calculated or extracted using standard formulas. The primary effect size was Pearson's correlation coefficient (r) which was used to specify the degree of relationship between emotional intelligence and the same outcome variables. Data extraction was carried out to guarantee comparability with an inter-rater agreement of over 90% (Fig. 1).

Emotional intelligence was evaluated through varied instruments (e.g., EQ-i WLEIS, MSCEIT, implicit EI coding, third-party behavioural measures) and outcome variables, such as trust in AI and user engagement, were evaluated by validated scales, behavioural measures, or self-reporting instruments. Publication partiality was evaluated by inspecting the symmetry of funnel plots and calculating the intercept from Egger's regression. These techniques help to detect potential lattices of missing small studies with non-significant results in the source data. Only investigations with a minimum quality edge were involved in the analysis. A random-effects model was employed to account for collection across research attitudes, sample types, and measurement modalities (Table 1).

4 Results

This meta-analysis combined evidence from 32 empirical studies with a total sample of 4720 North American, European, Asia Pacific, and Oceania participants. In all studies, the weighted mean effect size on the Fisher's z scale and N–3 weights was $r = 0.31$, 95% CI 0.28–0.33. This figure reflects a stable, moderate positive correlation between emotionally intelligent system design or perception and target socio-affective outcomes in human–AI interaction. The range of individual correlations is 0.24–0.48, ensuring the stability of the result. The most minor effects, $r = 0.24$, are

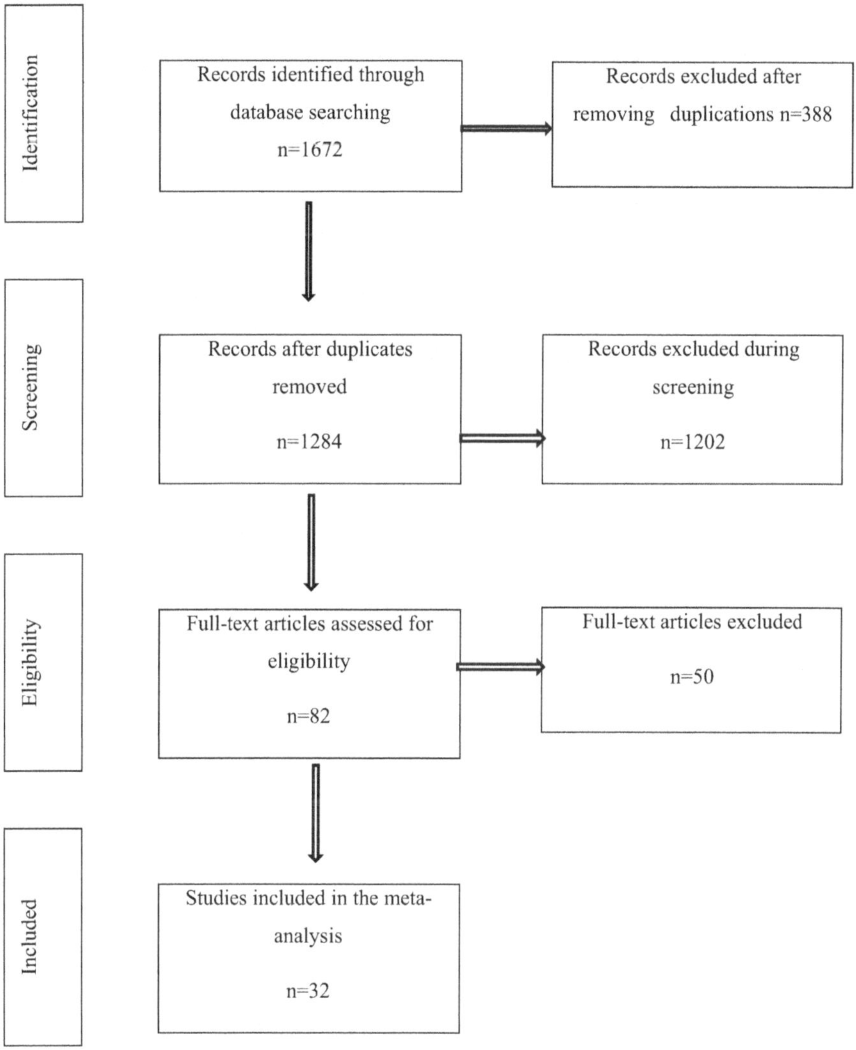

Fig. 1 PRISMA 2020 Flow diagram for study selection process. *Source* Authors' own compilation based on PRISMA (2020) guidelines

positive, while the largest, $r = 0.48$, suggests potential for significant gains when EI cues are salient. Moderator analysis discovered the employment of implicit emotional intelligence cues, including tonal variation and empathic postures. Anthropomorphic interfaces yielded slightly larger correlations ($r = 0.34$) than results derived from explicit trait-based self-report measures of emotional intelligence $r = .28$. The difference was not statistically significant $p > 0.05$ [77], so the two measures are

Table 1 Empirical studies on emotional intelligence in human-AI interaction (2000–2025)

Study ID	Authors	Country	N	AI system	EI measure	Outcome	*R*
1.	Ghandeharioun et al. [26]	USA	39	Chatbot	EI implicit	Engagement	0.34
2.	Ovsyannikova et al. [61]	Canada	556	AI-generated compassionate text vs human-generated text	Third-party ratings of compassion	Empathy	0.25
3.	Abdollahi et al. [1]	USA	16	Social robot	Custom implicit	Empathy	0.44
4.	Mathur et al. [49]	USA	31	Robot story teller	Empathic design	Empathy	0.39
5.	Nardelli et al. [56]	Italy	84	Social robot	Personality and emotion mapping	Trust	0.42
6.	Metcalf et al. [54]	UK	44	Voice assistant	Mirroring (implicit AI)	Trust	0.32
7.	Seymour and Van Kleek [82]	UK	60	Voice assistant	Anthropomorphic design (implicit EI)	Trust	0.33
8.	Cai et al. [8]	China	244	Conversational recommender	Trait-based self-report	Trust	0.29
9.	Obschonka et al. [59]	Germany	200	VR-embodied conversational agent (ECA) in virtual health	Implicit EI via similarity-matching	Engagement	0.28
10.	Hernandez et al. [35]	USA	120	Affective chatbot	Affective feedback (implicit EI)	Empathy	0.27
11.	Savery et al. [79]	USA	58	Music-prosody robot arm	Emotion prosody + gesture	Trust	0.28
12.	Pan et al. [63]	USA	200	GPT-4/Claude medical chatbots versus physicians	Empathy scores from blinded expert raters	Empathy	0.32
13.	Cheng et al. [12]	China	299	Text-based E-commerce chatbot	Trait EI (survey)	Trust	0.32

(continued)

Table 1 (continued)

Study ID	Authors	Country	N	AI system	EI measure	Outcome	*R*
14.	Mancone et al. [48]	Italy	300	Alexa voice assistant using empathic prosody	Implicit EI	Engagement	0.43
15.	Anzabi and Umemuro [3]	Norway	104	Social robot (pepper)	Listening-style adaptation	Trust	0.32
16.	de Rooij et al. [18]	Netherlands	110	Social robot (pepper) with facial/vocal mood expression	Implicit EI	Engagement	0.31
17.	Gu et al. [33]	China	462	Chatbot with perceived empathic ability	Trait-like EI perception	Trust	0.28
18.	Lee [41]	USA	76	Chatbot with anthropomorphic backstory	Implicit emotional framing	Engagement	0.27
19.	Yang and Xie [91]	USA	80	Social robot (appearance variation)	Emotional contagion via design cues	Empathy	0.31
20.	Chaudhry and Debi [10]	USA	159	Conversational agent (Wysa)	Implicit EI perceptions	Engagement	0.28
21.	Jolibois et al. [37]	Japan	70	Embodied conversational agent	Response timing affects impression	Empathy	0.27
22.	Saffaryazdi et al. [76]	NZ	40	Physiologically adaptive chatbot	Neural/physiological empathy mapping	Empathy	0.35
23.	Lim et al. [43]	China	343	E-commerce chatbot	Perceived interactivity & humanness (implicit EI)	Trust	0.31

(continued)

Table 1 (continued)

Study ID	Authors	Country	N	AI system	EI measure	Outcome	*R*
24.	Roesler et al. [75]	USA	312	Social robot	Trust attitude-behaviour alignment	Trust	0.29
25.	Savery et al. [80]	USA	48	Robotic arm with musical prosody	Affective prosody cues	Trust	0.33
26.	Ma et al. [47]	China	105	Emotion-aware voice assistant	Affective voice feedback	Engagement	0.30
27.	Cho et al. [14]	USA	40	Voice assistant (Alexa)	Backchanneling cues	Empathy	0.36
28.	Lin et al. [44]	USA	150	Mental-health chatbot	Empathy cues	Engagement	0.24
30.	Coker and Thakur [15]	Germany	48	Voice assistant	Vocal empathy cues	Trust	0.33
31.	Frederiksen et al. [23]	USA	220	Emotionally framed shopping assistant robot	Implicit EI via narrative empathy	Empathy	0.30
32.	Sanjeewa et al. [78]	Australia	57	Vocal empathy detection model using counselor call data	Vocal features predictive of counselor empathy	Empathy	0.48
33.	Chen et al. [11]	USA	45	Claude-based medical chatbot	Self-reported empathy ratings	Empathy	0.45

Source Authors' own compilation based on published empirical studies (2000–2025)
Notes N = sample size; R = Pearson's correlation coefficient; EI = Emotional Intelligence; AI = Artificial Intelligence

equally reasonable. Comparison between hardware-based and software-based platforms yielded no significant differences in effect sizes, so overall modality stability of the effect is supported.

Heterogeneity testing yielded Cochran's $Q = 16.22$ (31 df), $I^2 = 0\%$, suggesting most observed variance is explained by sampling error and the underlying relationship is highly stable in the face of variance of study context, AI modality, and measurement tradition. Corpus partitioning by outcome domain revealed highly convergent point estimates: Engagement ($k = 8$) $r = 0.32$ [0.27–0.37], Trust ($k = 12$) $r = 0.31$ [0.27–0.34], and Empathy ($k = 12$) $r = 0.30$ [0.25–0.34]. The engagement yielded numerically the most significant association, the overlapping confidence intervals demonstrate no statistically significant differentiation of the three outcomes, and emotional intelligence cues impart similarly positive effects across affective, relational, and attentional aspects of user experience. The publication bias analysis showed no asymmetry in the funnel plot, and Egger's regression test results were not statistically significant ($p = 0.28$) and had a low likelihood of publication bias. This result adds to the confidence in the consistency and validity of the meta-analytic estimate.

Descriptive inspection uncovered several consistencies. Experiments using implicit EI manipulations, such as empathic voice assistants [15]. Anthropomorphic backstories in chatbots typically reported correlations at or above the pooled mean, while self-report trait EI perceptions only produced slightly weaker but still positive effects [8]. Modality of intervention made no material difference: social robots. de Rooij et al. [18] Conversational agents, large language model chatbots of reduced convergent magnitudes, testifying to the generalisable nature of the finding across hardware and software instantiations [63].

Sample size did not systematically bias the meta-analytic estimate. Small-scale lab experiments $N = 16$ and large-scale field experiments $N = 556$ contributed proportionally to the weighted mean, and leave-one-out diagnostics changed the pooled r by no more than $\pm$ 0.02. The quantitative meta-analysis shows that artificial intelligence systems' presence or emotional intelligence signalling consistently increases user empathy, trust, and engagement. The low heterogeneity suggests some level of universality across geographic locations, interaction channels, and measurement methods, and emotionally intelligent functionality brings similar benefits. These findings provide a statistical benchmark against which to measure future design research and guidance for developers and policymakers who wish to make AI technologies more human-centred without sacrificing transparency and ethics.

5 Discussion

The present meta-analysis provides strong quantitative evidence for the socio-affective advantages of employing emotionally intelligent design features in artificial agents. Integrating data from 32 studies involving 4720 participants using chatbots, voice assistants, social robots, and immersive embodied agents, we found a medium-sized overall correlation (r = 0.31; 95% CI 0.28–0.33) across three primary outcomes: empathy, trust, and engagement across emotional intelligence measures. That all individual correlations were positive (r = 0.24–0.48) and heterogeneity was low ($I^2 \approx 0\%$) underscores the universality of this relationship across cultures, modalities, and measurement approaches.

The current results add to two decades of theory suggesting that people use social heuristics when media have human-like cues [57]. In keeping with the Computers Are Social Actors theory, users respond to emotionally expressive agents as intentional collaborators; even small cues like warmth of voice, mirroring, and emotive label feedback are sufficient to trigger empathic concern and modify interpersonal trust [34]. The equivalence of effect sizes across hardware modalities, robots vs. voice-only and software avatars chatbots, scripted embodied conversational agents, suggests that the simulation of emotional intelligence, not technical realization, drives user response. This result is consistent with social presence theory, which holds that warmth- and immediacy-related cues, not physical appearance, are the determinants of relational outcomes [4].

At the applied level, the very similar mutual correlations among empathy (r = 0.30), trust (r = 0.31), and engagement (r = 0.32) indicate that designers can implement an effectively integrated approach rather than optimizing each construct Implicit emotional intelligence manipulations, such as modulation, smiling eye light-emitting diodes, or backchannel acknowledgments, were typically superior to explicit trait EI self-report formats, guiding human-to-human research that emotions expressed through behaviour are more effective than explicit expressions [6]. Effect sizes were not artificially exaggerated in small sample laboratory experiments. Ethical and situation-specific issues temper these encouraging results. Since emotional expressions can be done without feeling, an empathetic response by a machine risks emotional deception if users ascribe moral agency to a device where it does not exist [64]. Additional studies indicated over-reliance on or anthropomorphic affection for machines by vulnerable groups. Designers should incorporate explicit artificial identity disclosures, already mandated by forthcoming regulations, the EU AI Act, and design opt-out mechanisms for emotionally learning functions. Cultural compatibility is another moderating issue: robots expressing high-intensity positive emotions were rated as insincere by collectivist East Asian samples, thus replicating previous research on display rule differences [50]. Policies addressing adaptive emotions that vary in expressiveness based on local culture could prevent such mismatches.

The measurement methods utilized are a methodological limitation. Although all studies used Pearson correlations, close to forty percent used single-item scales for empathy or trust, thus limiting construct validity. Future HAI studies may be assisted by multi-method triangulation combining behavioural logs, physiological recordings, and validated multi-item questionnaires to capture socio-affective experience's rich multidimensionality [9].

The limited range of existing research on short, scripted conversations and Western user populations. With emotionally intelligent agents coming into high-stakes environments, mental health therapy, geriatric care robots, autonomous cars, contextual stressors, user diversity, and prolonged exposure will introduce variability not encountered here. Effect size estimation should therefore be replicated in longitudinal, multi-site trials and under-represented demographic segments to evaluate generalisability.

Our synthesis guides policy and design practice. A pooled R of 0.31 across all experiments translates to roughly ten percent explained variance in user empathy, trust, or engagement, which is significant for a single design variable. Product teams should regard emotional expressiveness as one of the fundamental usability dimensions, alongside functionality and security. However, they should also implement safeguards: user-controllable emotion settings, trust recalibration reminder prompts at regular intervals, and bias audits on emotion detection pipelines. Regulators can take this evidence to design risk-tiered guidelines: systems for children, patients, or cognitively impaired users require more robust auditing of emotive tactics than cheerful customer service chatbots. This meta-analysis describes a significant, medium-strength association between emotional intelligence measures and positive socio-affective outcomes in human-artificial intelligence interactions.

6 Conclusion

This meta-analysis systematically synthesizes the emotional intelligence indication effect on human reaction to artificial agents. Through meta-analysis of 32 empirical studies across various platforms, ranging from chatbots and social robots to conversational voice assistants, the findings establish a strong and moderate positive association between emotionally intelligent design and three primary user outcomes: empathy, trust, and engagement. With a $r = 0.31$ correlation, with low heterogeneity, the results show that emotional responsiveness can be measured and replicated across contexts and modalities of interaction. It is consistent with current debates about machine social understanding and proves that socially and emotionally relevant cues can genuinely motivate social and emotional reactions when implemented appropriately. Whether through empathetic tone, responsive facial affect, or fitting backchannel signals, these cues establish social relationships and ensure that user's attention stays on the robot and that positive emotional outcomes are achieved. The convergence of findings in empathy, trust, and engagement implies that designers do not need to do everything to make the cues socially and emotionally relevant. Strategies of emotionally intelligent communication can achieve all three outcomes at once.

Methodologically, the results of this study provide developers and regulators with a solid empirical foundation for making choices in design, regulation, and ethical protection in the future. As emotional intelligence systems become pervasive in parts such as healthcare, tutoring, and customer service to understand their limits and best practices in design. Most of the studies considered are short-term and laboratory-based, with limited cultural diversity and comparatively simple interaction tasks. There is an enormous need for cross-cultural and longitudinal validation and for the triangulation of humans' subjective and behavioral trust, empathy, and engagement responses. The challenge for researchers and developers will not be just to make machines more emotionally intelligent but in humanly meaningful ways, adaptive to user agency, and sensitive to cultural, ethical, and contextual nuance. Emotional intelligence in AI is not a technical feat but a relational perspective that, if pursued thoughtfully, honourably, and boldly, could be extraordinarily powerful.

Declaration of Generative AI and AI-Assisted Technologies in the Chapter Preparation Process During the preparation of this chapter, the author (s) used Microsoft Excel for organizing, screening, and managing bibliographic records and meta-analytic data, including calculation and verification of effect sizes. In addition, Grammarly was used as an AI-assisted language tool to support grammar checking and improve clarity and readability of the manuscript. After using these tools, the author (s) carefully reviewed, edited, and validated all content and take full responsibility for the accuracy, originality, and integrity of the chapter.

References

1. Abdollahi, H., Mahoor, M.H., Zandie, R., Siewierski, J., Qualls, S.H.: Artificial emotional intelligence in socially assistive robots for older adults: a pilot study. IEEE Trans. Affect. Comput. **14**(3), 2020–2032 (2022)
2. Al-Shafei, M.: Navigating human-chatbot interactions: an investigation into factors influencing user satisfaction and engagement. Int. J. Hum.–Comput. Interact. **41**(1), 411–428 (2025)
3. Anzabi, N., Umemuro, H.: Effect of different listening behaviors of social robots on perceived trust in human-robot interactions. Int. J. Soc. Robot. **15**(6), 931–951 (2023)
4. Biocca, F., Harms, C., Burgoon, J.K.: Toward a more robust theory and measure of social presence: review and suggested criteria. Presence: Teleoperators Virtual Environ. **12**(5), 456–480 (2003)
5. Boyatzis, R.E., Goleman, D., Rhee, K.: Clustering competence in emotional intelligence: insights from the emotional competence inventory (ECI). Handb. Emot. Intell. **99**(6), 343–362 (2000)
6. Brave, S., Nass, C., Hutchinson, K.: Computers that care: investigating the effects of orientation of emotion exhibited by an embodied computer agent. Int. J. Hum.-Comput. Stud. **62**(2), 161–178 (2005)
7. Broadbent, J., Panadero, E., Boud, D.: Implementing summative assessment with a formative flavour: a case study in a large class. Assess. Eval. High. Educ. **43**(2), 307–322 (2018)
8. Cai, W., Jin, Y., Chen, L.: Impacts of personal characteristics on user trust in conversational recommender systems. In: Proceedings of the 2022 CHI Conference on Human Factors in Computing Systems, pp. 1–14 (2022)
9. Calvo, R.A., D'Mello, S.: Affect detection.: an interdisciplinary review of models, methods, and their applications. IEEE Trans. Affect. Comput. **1**(1), 18–37 (2010)
10. Chaudhry, B.M., Debi, H.R.: User perceptions and experiences of an AI-driven conversational agent for mental health support. Mhealth. **10**, 22 (2024)
11. Chen, D., Chauhan, K., Parsa, R., Liu, Z.A., Liu, F.-F., Mak, E., Eng, L., Hannon, B.L., Croke, J., Hope, A.: Patient perceptions of empathy in physician and artificial intelligence chatbot responses to patient questions about cancer. Npj Digit. Med. **8**(1), 1–5 (2025)
12. Cheng, X., Bao, Y., Zarifis, A., Gong, W., Mou, J.: Exploring consumers' response to text-based chatbots in e-commerce: the moderating role of task complexity and chatbot disclosure. Internet Res. **32**(2), 496–517 (2021)
13. Chi, N.T.K., Hoang, V., N.: Investigating the customer trust in artificial intelligence: the role of anthropomorphism, empathy response, and interaction. CAAI Trans. Intell. Technol. **8**(1), 260–273 (2023)

14. Cho, E., Motalebi, N., Sundar, S.S., Abdullah, S.: Alexa as an active listener: how backchanneling can elicit self-disclosure and promote user experience. Proc. ACM Hum.-Comput. Interact. **6**(CSCW2), 1–23 (2022)
15. Coker, K.K., Thakur, R.: Alexa, may I adopt you? The role of voice assistant empathy and user-perceived risk in customer service delivery. J. Serv. Mark. **38**(3), 301–311 (2023)
16. Crawford, K.: The Atlas of AI: Power, Politics, and the Planetary Costs of Artificial Intelligence. Yale University Press (2021)
17. Davis, M.H.: The effects of dispositional empathy on emotional reactions and helping: a multidimensional approach. J. Pers. **51**(2), 167–184 (1983)
18. de Rooij, A., Broek, S. van den, Bouw, M., & de Wit, J.: Co-creating with a robot facilitator: robot expressions cause mood contagion enhancing collaboration, satisfaction, and performance. Int. J. Soc. Robot., 16 (11), 2133–2152 (2024)
19. De Visser, E.J., Monfort, S.S., McKendrick, R., Smith, M.A.B., McKnight, P.E., Krueger, F., Parasuraman, R.: Almost human: anthropomorphism increases trust resilience in cognitive agents. J. Exp. Psychol. Appl. **22**(3), 331 (2016)
20. Devaram, S.: Empathic Chatbot: Emotional Intelligence for Empathic Chatbot: Emotional Intelligence for Mental Health Well-Being. ArXiv Preprint ArXiv:2012.09130 (2020)
21. Erol, B.A., Majumdar, A., Benavidez, P., Rad, P., Choo, K.-K.R., Jamshidi, M.: Toward artificial emotional intelligence for cooperative social human–machine interaction. IEEE Trans. Comput. Soc. Syst. **7**(1), 234–246 (2019)
22. Fitzpatrick, K.K., Darcy, A., Vierhile, M.: Delivering cognitive behavior therapy to young adults with symptoms of depression and anxiety using a fully automated conversational agent (Woebot): a randomized controlled trial. JMIR Ment. Health. **4**(2), e7785 (2017)
23. Frederiksen, M.R., Fischer, K., Matarić, M.: Robot vulnerability and the elicitation of user empathy. In: 2022 31st IEEE International Conference on Robot and Human Interactive Communication (RO-MAN), pp. 52–58 (2022)
24. Frühbeis, C., Fröhlich, D., Krämer-Albers, E.-M.: Emerging roles of exosomes in neuron–glia communication. Front. Physiol. **3**, 119 (2012)
25. Fulmer, S.M., D'Mello, S.K., Strain, A., Graesser, A.C.: Interest-based text preference moderates the effect of text difficulty on engagement and learning. Contemp. Educ. Psychol. **41**, 98–110 (2015)
26. Ghandeharioun, A., McDuff, D., Czerwinski, M., Rowan, K.: EMMA.: an emotion-aware wellbeing chatbot. In: 2019 8th International Conference on Affective Computing and Intelligent Interaction (ACII), pp. 1–7 (2019)
27. Gillath, O., Ai, T., Branicky, M.S., Keshmiri, S., Davison, R.B., Spaulding, R.: Attachment and trust in artificial intelligence. Comput. Hum. Behav. **115**, 106607 (2021)
28. Glikson, E., Woolley, A.W.: Human trust in artificial intelligence: review of empirical research. Acad. Manag. Ann. **14**(2), 627–660 (2020)
29. Gnewuch, U., Morana, S., Adam, M., & Maedche, A.: Faster Is Not Always Better: Understanding the Effect of Dynamic Response Delays in Human-Chatbot Interaction (2018)
30. Gnewuch, U., Morana, S., Maedche, A.: Towards designing cooperative and social conversational agents for customer service. ICIS. **1–13** (2017)
31. Goleman, D., Intelligence, E.: Why it can matter more than IQ. Emot. Intell. (1995)
32. Gratch, J., Marsella, S.: A domain-independent framework for modeling emotion. Cogn. Syst. Res. **5**(4), 269–306 (2004)
33. Gu, C., Zhang, Y., Zeng, L.: Exploring the mechanism of sustained consumer trust in AI chatbots after service failures: a perspective based on attribution and CASA theories. Humanities Soc. Sci. Commun. **11**(1), 1–12 (2024)
34. Hancock, P.A., Billings, D.R., Schaefer, K.E., Chen, J.Y.C., De Visser, E.J., Parasuraman, R.: A meta-analysis of factors affecting trust in human-robot interaction. Hum. Factors. **53**(5), 517–527 (2011)
35. Hernandez, J., Suh, J., Amores, J., Rowan, K., Ramos, G., & Czerwinski, M.: Affective Conversational Agents: Understanding Expectations and Personal Influences. ArXiv Preprint ArXiv:2310.12459 (2023)

36. Jiang, Q., Zhang, Y., Pian, W.: Chatbot as an emergency exist: mediated empathy for resilience via human-AI interaction during the COVID-19 pandemic. Inf. Process. Manag. **59**(6), 103074 (2022)
37. Jolibois, S.C., Ito, A., Nose, T.: The development of an emotional embodied conversational agent and the evaluation of the effect of response delay on user impression. Appl. Sci. **15**(8), 4256 (2025)
38. Joseph, D.L., Newman, D.A.: Emotional intelligence: an integrative meta-analysis and cascading model. J. Appl. Psychol. **95**(1), 54 (2010)
39. Kim, S., Chen, J., Cheng, T., Gindulyte, A., He, J., He, S., Li, Q., Shoemaker, B.A., Thiessen, P.A., Yu, B.: PubChem in new data content and improved web interfaces. Nucleic Acids Res. **49**(D1), D1388–D1395 (2021)
40. Lee, J.D., See, K.A.: Trust in automation: designing for appropriate reliance. Hum. Factors. **46**(1), 50–80 (2004)
41. Lee, W.-J.: Service failure by human, service recovery by AI Chatbot: the impact of justice, AI efficacy on recovery effort. Virtual Econ. **7**(4), 48–63 (2024)
42. Leite, I., Pereira, A., Mascarenhas, S., Martinho, C., Prada, R., Paiva, A.: The influence of empathy in human–robot relations. Int. J. Hum.-Comput. Stud. **71**(3), 250–260 (2013)
43. Lim, S., Schmälzle, R., Bente, G.: Artificial social influence via human-embodied AI agent interaction in immersive virtual reality (VR): effects of similarity-matching during health conversations. Comput. Hum. Behav.: Artif. Humans. (2025)
44. Lin, S., Lin, L., Hou, C., Chen, B., Li, J., Ni, S.: Empathy-based communication framework for chatbots: a mental health chatbot application and evaluation. In: Proceedings of the 11th International Conference on Human-Agent Interaction, pp. 264–272 (2023)
45. Liu-Thompkins, Y., Okazaki, S., Li, H.: Artificial empathy in marketing interactions: bridging the human-AI gap in affective and social customer experience. J. Acad. Mark. Sci. **50**(6), 1198–1218 (2022)
46. Ma, X.: Towards human-engaged AI. In: *IJCAI*, pp. 5682–5686 (2018)
47. Ma, Y., Zhang, Y., Bachinski, M., Fjeld, M.: Emotion-aware voice assistants: design, implementation, and preliminary insights. In: Proceedings of the Eleventh International Symposium of Chinese CHI, pp. 527–532 (2023)
48. Mancone, S., Diotaiuti, P., Valente, G., Corrado, S., Bellizzi, F., Vilarino, G.T., Andrade, A.: The use of voice assistant for psychological assessment elicits empathy and engagement while maintaining good psychometric properties. Behav. Sci. **13**(7), 550 (2023)
49. Mathur, L., Spitale, M., Xi, H., Li, J., Matarić, M.J.: Modeling user empathy elicited by a robot storyteller. In: 2021 9th International Conference on Affective Computing and Intelligent Interaction (ACII), pp. 1–8 (2021)
50. Matsumoto, D., Yoo, S.H.: Toward a new generation of cross-cultural research. Perspect. Psychol. Sci. **1**(3), 234–250 (2006)
51. Mayer, J.D., Salovey, P.: Mayer-Salovery-Caruso Emotional Intelligence Test. Multi-Health Systems Incorporated Toronto (2007)
52. McStay, A.: Emotional AI and EdTech: serving the public good? Learn. Media Technol. **45**(3), 270–283 (2020)
53. Mesquita, B., Walker, R.: Cultural differences in emotions: a context for interpreting emotional experiences. Behav. Res. Ther. **41**(7), 777–793 (2003)
54. Metcalf, K., Theobald, B.-J., Weinberg, G., Lee, R., Jonsson, I.-M., Webb, R., & Apostoloff, N.: Mirroring to Build Trust in Digital Assistants. ArXiv Preprint ArXiv:1904.01664 (2019)
55. Mikolajczak, M.: Going beyond the ability-trait debate: the three-level model of emotional intelligence. E-J. Appl. Psychol. **5**(2), 25–31 (2009)
56. Nardelli, A., Sgorbissa, A., & Recchiuto, C. T.: Designing Empathetic Companions: Exploring Personality, Emotion, and Trust in Social Robots. ArXiv Preprint ArXiv:2504.13964 (2025)
57. Nass, C., Moon, Y.: Machines and mindlessness: social responses to computers. J. Soc. Issues. **56**(1), 81–103 (2000)
58. Nomura, T., Kanda, T., Suzuki, T., Kato, K.: Prediction of human behavior in human—robot interaction using psychological scales for anxiety and negative attitudes toward robots. IEEE Trans. Robot. **24**(2), 442–451 (2008)

59. Obschonka, M., Silbereisen, R.K., Schmitt-Rodermund, E.: Explaining entrepreneurial behavior: dispositional personality traits, growth of personal entrepreneurial resources, and business idea generation. Career Dev. Q. **60**(2), 178–190 (2012)
60. Oksanen, A., Savela, N., Latikka, R., Koivula, A.: Trust toward robots and artificial intelligence: an experimental approach to human–technology interactions online. Front. Psychol. **11**, 568256 (2020)
61. Ovsyannikova, D., de Mello, V.O., Inzlicht, M.: Third-party evaluators perceive AI as more compassionate than expert humans. Commun. Psychol. **3**(1), 4 (2025)
62. Page, M.J., McKenzie, J.E., Bossuyt, P.M., Boutron, I., Hoffmann, T.C., Mulrow, C.D., Shamseer, L., Tetzlaff, J.M., Akl, E.A., Brennan, S.E.: The PRISMA 2020 statement: an updated guideline for reporting systematic reviews. BMJ. **372** (2021)
63. Pan, A., Musheyev, D., Bockelman, D., Loeb, S., Kabarriti, A.E.: Assessment of artificial intelligence chatbot responses to top searched queries about cancer. JAMA Oncol. **9**(10), 1437–1440 (2023)
64. Parviainen, J., Coeckelbergh, M.: The political choreography of the Sophia robot: beyond robot rights and citizenship to political performances for the social robotics market. AI Soc. **36**(3), 715–724 (2021)
65. Pataranutaporn, P., Liu, R., Finn, E., Maes, P.: Influencing human–AI interaction by priming beliefs about AI can increase perceived trustworthiness, empathy and effectiveness. Nat. Mach. Intell. **5**(10), 1076–1086 (2023)
66. Pelau, C., Dabija, D.-C., Ene, I.: What makes an AI device human-like? The role of interaction quality, empathy and perceived psychological anthropomorphic characteristics in the acceptance of artificial intelligence in the service industry. Comput. Hum. Behav. **122**, 106855 (2021)
67. Petrides, K.V., Mikolajczak, M., Mavroveli, S., Sanchez-Ruiz, M.-J., Furnham, A., Pérez-González, J.-C.: Developments in trait emotional intelligence research. Emot. Rev. **8**(4), 335–341 (2016)
68. Petrides, K.V., Pita, R., Kokkinaki, F.: The location of trait emotional intelligence in personality factor space. Br. J. Psychol. **98**(2), 273–289 (2007)
69. Picard, R.W.: A Ective Computing. MIT press Cambridge Massachusetts (1997)
70. Pitardi, V., Marriott, H.R.: Alexa, she's not human but… unveiling the drivers of consumers' trust in voice-based artificial intelligence. Psychol. Mark. **38**(4), 626–642 (2021)
71. Poole, K.L., Schmidt, L.A.: While a shy child waits: autonomic and affective responses during the anticipation and delivery of a speech. Emotion. **21**(5), 1103 (2021)
72. Prentice, C., Nguyen, M.: Engaging and retaining customers with AI and employee service. J. Retail. Consum. Serv. **56**, 102186 (2020)
73. Rad, H.S., Alipour, R., Jafarpour, A.: Using artificial intelligence to foster students' writing feedback literacy, engagement, and outcome: a case of Wordtune application. Interact. Learn. Environ. **32**(9), 5020–5040 (2024)
74. Rehm, M., André, E.: Catch me if you can: exploring lying agents in social settings. In: Proceedings of the Fourth International Joint Conference on Autonomous Agents and Multiagent Systems, pp. 937–944 (2005)
75. Roesler, E., Vollmann, M., Manzey, D., Onnasch, L.: The dynamics of human–robot trust attitude and behavior—exploring the effects of anthropomorphism and type of failure. Comput. Hum. Behav. **150**, 108008 (2024)
76. Saffaryazdi, N., Gunasekaran, T. S., Laveys, K., Broadbent, E., & Billinghurst, M.: Empathetic Conversational Agents: Utilizing Neural and Physiological Signals for Enhanced Empathetic Interactions. ArXiv Preprint ArXiv:2501.08393 (2025)
77. Sánchez-Álvarez, N., Berrios Martos, M.P., Extremera, N.: A meta-analysis of the relationship between emotional intelligence and academic performance in secondary education: a multi-stream comparison. Front. Psychol. **11**, 1517 (2020)
78. Sanjeewa, R., Iyer, R., Apputhurai, P., Wickramasinghe, N., Meyer, D.: Machine learning approach to identifying empathy using the vocals of mental health helpline Counselors: algorithm development and validation. JMIR Form. Res. **9**, e67835 (2025)

to invent the media without losing journalistic integrity and democratic values; to be more specific, eventually, AI-backed journalism will become a tool.

Keywords Artificial intelligence in journalism · Automated news generation · Media ethics and AI · Algorithmic bias in news · AI-driven audience engagement

1 Introduction

At crossroads of theoretical technological changes are journalism in the fast-growing digital world that is modifying classical practices. AI is a needle wound, and it basically transforms the content of production, distribution, and consumption of news [1]. The world has changed journalistic procedures through AI embracing the possibilities of machine learning, natural language processing (NLP), data mining, and automation over the world with significant opportunities and considerable challenges [2, 3]. Technology has reduced the reliability, impartiality, and effectiveness of the media, and it now establishes devious moral challenges of authenticity, prejudice, openness, and responsibility, particularly in moments as individuals begin to lose faith in the information channels [4].

The use of AI in journalism is vast, where the use of AI applications is automating the generation of information, giving personalized coverage, real-time fact checking, and mood analysts of the audience. The large media companies such as the Associated Press, The Washington Post, Reuters, and Bloomberg have adopted the technology of developing standard reports on issues such as finance, sports, and news on election, via the Heliograph and Cyborg [5, 6]. These systems generate a vast amount of content from virtually nothing in no time, in different languages and even in real time and have relieved much of the pressure that had traditionally befallen human reporters. Due to this fact, AI has been hailed as a way of enhancing productivity, scalability, and efficiency in newsrooms [7, 8]. This technological move is, however, a serious concern. The current issue is that it may conflict with editorial authority and journalistic independence as hinting at approval using algorithms will probably destroy the fine setting, the richness of being narratives, and the critical thinking that only human journalists could provide [9]. Furthermore, many AI systems are presented as black boxes, amorphous and not explaining how they have reached the certain answer but presenting it to a person, hence it is difficult to see whether information provided by AI systems is true and because of that, people do not believe the media [10].

The next important issue is the problem of algorithmic bias and its contribution to misinformation. In case the data sets used to train AI are biased against society, the produced systems will either enhance stereotyping, discriminate against minorities, or misrepresent reality [11, 12]. Furthermore, the current developments in the domain of deepfakes and synthetic media make it possible to create fake content and obscure the line between reality and fake information, which presents a serious risk to the reputability of journalism and the stability of democracy in general, particularly on the situation with elections [13, 14].

More positively, AI greatly helps to increase engagement and individualization of the content to the audience. Through user behavior analysis, preference, and interaction patterns, platforms could provide users with personalized news, increasing user retention and satisfaction [15, 16]. Chatbots and virtual assistants based on AI principles allow interacting in real time, allowing the user to inquire about information, provide a summary, or further explore the subject, which is associated with the current demands when content needs to be evaluated on demand and be relevant at the same time [17].

However, personalization has also negative aspects. It can also be used to create filter bubbles or echo chambers where algorithms can support previous beliefs while limiting exposure to other ideas and impairing the capacity of journalism to develop a discourse about information, informed, and balanced in the general population [18]. Therefore, AI systems must be molded, more captivating, and less varied in outlook.

The problem of labor, skills acquisition, and professional transformation can also be considered with the aid of AI application in journalism. AI will ease the duties associated with repetitive and data-intensive work, it is not able to replace human journalists in whatever regards investigative reporting and making ethical decisions or accomplishing work with the aim in accountability. It means that there should be a working model where the AI is a thing of completion rather than replacement of the journalists, digital literacy, algorithmic knowledge, and data processing should be trained, and corporations should invest on ethical AI systems [19].

The influence of AI is rather strong in such a context as the rapidly digitalizing democracy of India, which is shaped by the news consumption that relies on mobile devices, increasing digital populations, and the popularity of smartphones. Indian newsrooms use AI to facilitate multilingual translations and regional content creation and social media analytics to get better access to information in under-served regions [20]. The loose laws, however, just increase the likelihood of abuse, and there are no more prospects than an ethical implementation of AI, strong regulatory customs, and comforts to repositioning the US government against business on this point [21].

Additionally, the lack of variety in the media and the concentration of AI technology to various players are threats to the independent existence and variety of the media. Such companies as Google and Meta can access the priorities and main data pipelines that define visibility and priority [22]. Rules are crucial in democratizing the right of access to AI and ensuring that power over dialogue is unreasonable.

AI is a powerful enabler and a potential intruder of journalism. It is concerned with inefficiencies in style of content creation and interaction with the viewer and poses moral hazard and dangers to democracy. The outlook of journalism must be that involving the combination of the AI and human regulation, protectionist regulations, and the deep conviction of the values of journalism, which means that technology will never substitute but assist the search of the truth, accountability, and serving of the population [23].

Objectives

1. To examine the transformative impact of AI technologies on news production and editorial workflows in journalism.
2. To evaluate the ethical challenges posed by AI in journalism, focusing on misinformation, bias, and transparency.
3. To propose evidence-based policy recommendations for the ethical and sustainable integration of AI in journalism.

2 Review of Literature

The adoption of AI in journalism has caused a strong and growing literature of chapter, including academic articles, industrial reviews, and cross-disciplinary research. This literature discusses the revolutionary effects of AI on news production, ethical aspects, editorial processes and audience interaction, its possibilities, and challenges [24]. Recent systematic reviews draw attention to the trends in the adoption of AI worldwide, such as regulatory problems and ethical issues, which highlights the importance of critical AI literacy in journalists when working in the changing environment [25]. Much of the literature is dedicated to automated content generation by natural language generation (NLG) systems that can produce routine news stories on a faster basis. Graefe et al. have examined the degree to which readers trust automated news, as they are considered to be efficient and credible, yet in comparison with the news that is developed by human authors, they tend to be less emotional and narrative-filled [26]. In their vision, Marconi and Siegman envision a form of augmented journalism, in which the AI tools can take over many data-heavy tasks in newsrooms to enable the reporter to concentrate on interpreting the information [27]. Other recent works also explore the use of AI that generates scalable content, pointing to the threat of homogeneous reporting on all platforms, and the potential to require editorial supervision to keep diversity [28].

The concept of ethical implications is among the primary themes, in terms of algorithmic transparency. Diakopoulos maintains that black box systems are opaque and, hence, portend no accountability, because journalists find it hard to question the outputs of computations, making it more difficult to trust in AI-generated news [29]. Carlson discusses how AI transfers journalistic power to algorithms depriving human professionals of control over the process, changing the nature of power in the newsroom [30]. The advanced research suggests that hybrid human–machine communication systems can be used to balance computational accuracy and human judgment, so that AI-assisted journalism could be trusted [31].

The discussion on ethics is also applied to synthetic media where Chesney and Citron address the risks of deepfakes in controlling the discourse of the population and endangering democratic institutions by erasing the boundary between reality and fiction [32]. Rini proposes the idea of strengthening the verification mechanism to regain epistemic trust and warns that the widespread deepfakes can lead to the development of general mistrust in the media [33]. Recent research on AI-driven fact-checking has shown better success in fighting misinformation yet it cautions against over-reliance because biases in the training data can carry over inaccuracy [34].

Personalization which is aided by AI has been questioned as divided audiences. Helberger criticizes recommendation systems based on the effects of forming echo chambers, where people will not be exposed to different opinions and pluralism in media consumption will be undermined [35]. The presence of filter bubbles is a contentious issue that Bodo et al. discuss in a mixed manner, yet the authors state that ideological silos are supported by algorithmic curation in many instances [36]. The legendary idea of the filter bubble by Pariser is still commonly used, and the latest studies use this concept to the new AI-based news feeds that are not as informative as they are engaging [37]. More sophisticated proposals imply the creation of AI systems that would encourage serendipity and inter-ideological exposure to prevent polarization [38].

Another important area is the engagement of the audience using AI products, including chatbots and sentiment analysis. Beckett advocates hybrid forms of journalism were artificial intelligence supplements human editorial judgment, without damaging ethical principles or investigative journalism [39]. Nevertheless, the fear of simplistic material that appeals to personal tastes is still there and it can reduce the degree of journalism [40].

3 Research Methodology

3.1 Research Design

This research chapter uses a qualitative research method, which is descriptive, to discuss how AI technologies transform the news production and editorial processes, assess the ethical issues raised by them, and offer policy implications regarding the role of ethical integration in journalism. The method is quite appropriate in synthesizing secondary data of emerging phenomena and giving detailed descriptions and explanations without having to empirically test it. With the emphasis on secondary analysis, the design will enable conducting a thorough overview of the available literature, case studies, and policy documents and identify patterns, trends, and implications of the application of AI to journalism.

3.2 Data Collection Methods

The secondary sources will be used exclusively to collect the data, such as peer-reviewed academic articles, industry reports, white chapters, case studies, and policy frameworks released in 2015–2025. These sources were found on the academic databases like Google Scholar, JSTOR, and PubMed and the organizational repositories of institutions like UNESCO and Reuters Institute. The period reflects the swift development of AI technologies since the beginning of their use to the ubiquitous presence in the media practices.

In Objective 1 (understanding how AI is changing the nature of the news production and workflow), the information consists of case studies of such tools as Heliograf (The Washington Post) and automated systems in Reuters, based on the industry analysis and critical review of the literature.

Regarding Objective 2 (assessing ethical issues in misinformation, bias, and transparency), the scholarly articles and reports on algorithmic biases, deepfakes, and ethical concerns in AI-generated content were used.

In Objective 3 (advocating policy recommendations), since the issue involves policy proposals, information is obtained based on the ethical theories and policy documents, including AI4People and UNESCO guidelines, focusing on the sustainable and responsible use of AI in journalism.

3.3 Analytical Framework

The data were organized and interpreted using thematic analysis, through which there was the need of repeated coding in order to derive recurring themes like automation efficiencies, ethical risks and governance needs. NVivo software helped to code the content and divide it into themes that corresponded with the stated objectives: workflow changes, ethical risks (e.g., amplification of bias), and policy interventions. This framework relies on the theories of media ethics to put the findings into perspective so as to have a balanced evaluation of the advantages and disadvantages of AI.

3.4 Case Selection

The cases were selected intentionally to reflect various uses of AI in journalism. Western cases involve the automated reporting of The Washington Post, fact-checking with AI of Reuters and Indian cases involve multilingual coverage using AI by The Quint and personalization algorithms by Times Internet. The selection criteria were focused on high impact implementations, geographic differences, and suitability to ethical and workflow themes, where comparative information about differences between the world and region might be obtained.

3.5 Scope and Delimitations

The chapter is restricted to the analysis of secondary data wherein no primary data collection methods such as surveys or interviews are to be employed in the chapter to avoid loss of direction to the descriptive synthesis.

Time frame: Not older than 2015–2025 since this is a current area of AI.

Geographic Scope: Global, but with a focus on United States and India to present the differences in developed and emerging media ecosystem.

3.6 Limitations

Being a secondary analysis, the chapter does not have primary data, which could be lacking in depth and real-time data of the practitioners. Proprietary AI algorithms are not easily accessible, and these may topicalize transparency discourse. Due to the rapid development of AI, some of the findings can become outdated after 2025. Nonetheless, these problems are resolved by methodological rigor using source triangulation and thematic validation which gives a sound basis to the analysis.

4 Findings

This chapter is based on the synthesis of secondary literature with identification of the essential trends, challenges, and recommendations concerning AI in journalism. They are structured based on the research objectives and take the recent academic, industry, and policy literature to develop a descriptive outline.

4.1 Transformative Impact on News Production and Editorial Workflows

AI technologies have significantly transformed the way news is produced by automating the repetitive processes, increasing the efficiency, and permitting novel ways of news production. In larger ones, like the Heliograf system at The Washington Post, reporting all the information-heavy topics like elections and sports is automated by the system, resulting in thousands of stories annually with hardly any human intervention, hence allowing newschapter reporters to perform more investigative journalism. Moreover, Bloomberg Cyborg is adopting the use of AI to assist in financial reporting, which integrates human control with AI and escalates the complex analysis of data quickly.

Table 1 AI tools used in news production

AI tool	Media organization	Primary function	Impact on workflows
Heliograf	The Washington post	Automated reporting on data-driven stories	Increases speed and volume of content production, reduces manual effort
Cyborg	Bloomberg	Assistance in financial news analysis	Enhances accuracy in complex data handling, supports hybrid human-AI editing
AI systems	Reuters	News gathering and dissemination	Enables real-time personalization and multilingual output
Generative AI	Tencent	Content generation and adaptation	Scales production for diverse audiences integrates with distribution channels

It finds applications in the editorial processes and uses AI to make delivery of the media across platforms easier. These media outlets such as Reuters and Tencent apply AI in the creation and publication of news, the creation of news collections, and multilingual adaptations and personalization of news through live news coverage. National news houses in Denmark have been discussing how to apply AI to simplify the editorial process (story generation, publication, and the like), which would eventually lead to the so-called smart newsrooms (that would partially rely on machine learning to provide predictive analytics and editorial data). However, such transformation shifts the roles of journalism in which AI is to collect data and verify their facts when editorial autonomy is required not to hinder the quality.

Table 1 provides a brief description of key AI tools, deployed in large newsrooms, and showing what they do as well as how they change the workflow. It demonstrates that AI is effective (in terms of speed and scalability) but not lose data-intensive tasks. The examples build on the practices existing around the world and they emphasize the role which will be played by technology and will alter the existing processes of the editorial though not to fully replace the human element.

Overall, the emergence of AI may result in innovation, but it must be approached carefully, otherwise they can flatten journalism.

4.2 Ethical Challenges Posed by AI in Journalism

The article focused on AI and journalism, highlighting potential ethical issues that may arise when humans are replaced in the workplace by automatons. The nature of AI poses a significant ethical dilemma in the journalistic field especially when it comes to fake news and biasness, and transparency. Among the main concerns may be such aspects as the spread of AI-generated content, main of which is deepfakes, which can be used to dominate discourse and lower the credibility of the media.

Studies indicate that when given biased data, trained AI systems produce further reinforcements of stereotypes and misinformation that contain cases whereby the generative AI generates defamation or a hoax. As an example, hate speech amplification and disinformation, which are connected to AI, was a phenomenon under the digital ecosystem of Indonesia, which proves that efficient ethical governance is essential.

Black box algorithms bring about transparency issues since in the absence of transparency in the decision-making process, no one is held to account in case the decision made harms others. The above confusion raises the question of how to point the blame to either the developers, journalists, or organisations in which misinformation or bias is created by using AI to make them. Besides, Global South is grappling with disparities in accessibility and systems of ethics, which are likely to widen digital inequalities.

4.3 Policy Recommendations for Ethical and Sustainable Integration

To counter such issues, evidence-based policies refer to transparency, accountability, and human control. AI policies should be implemented in newsrooms to alleviate the threat of disseminating misinformation such as misinformation reporting and maintenance of fair access. Such approaches as the application of antitrust policies to achieve the integrity of information and a competitive tone, as well as such guidelines as AI4People, are recommended on fairness and sustainability. Frameworks provided by UNESCO and the Council of Europe design human rights-focused AI applications, and countries deploy national strategies that aim to train and perform an ethical audit.

To make AI integration sustainable, the perceived environmental impact of this technology should be reduced as much as possible, and more equitable methods of hiring need to be encouraged. It can be seen in organizations such as Schibsted whereby there is an illuminated rule and human compliance.

5 Discussion

The results obtained within the frames of this research are crucial in highlighting the complex nature of AI application in journalism in case of being compared with the general academic and business debate concerning the topic of technological discontinuity in media ecosystems. Analyzing the transformative effect of AI on news creation and editorial processes (Objective 1), one can refer to evidence that the since automation tools may increase efficiency and scalability via the provided example of Heliograf and Cyborg, which are the data-driven reporting systems where reporting is automated, and the journalist can concentrate on more useful activities.

This can be associated with institutional views on generative AI, which consider it a disruptive power reorganizing the newsroom practice by reallocating its decision-making data and workflow streamlining. This change, however, also undermines some of the traditional editorial authority and may threaten to substitute the human judgment with an algorithmic one especially in resource constrained contexts such as on the Global South. The implications include the fact that the development of AI helps create innovation: e.g., smart newsrooms are using predictive analytics; however, corresponding adaptive measures will need to be implemented in order to maintain journalistic autonomy and respond to new roles.

In terms of the problems of ethics (Objective 2), the focus of the chapter, namely, misinformation, prejudice, and transparency resonate with the threat of AI and its capability to enhance fake news and biased information search that results to discriminatory or misleading information. As an example, deepfakes and AI-based misinformation presents a threat to the social confidence of the populace and democracy, which is reflected in skewed datasets sustaining harm in analysis. More so, it agrees with the UNESCO ethical framework that alerts AI duplicating the biases and discrimination in the real world, unless proper measures are taken. This makes accountability difficult as well since black box systems may not be very transparent in its operations making the issue of liability in defamatory or harmful produce of journalists a tough one. Such issues are especially pressing when it comes to fighting misinformation, and fact-checking AI tools should be confined in terms of possible ethical breaches. The discourse presupposes the greater intensity of the AI-literacy of journalists to reduce the occurrence of these risks so that the trend of technological adoption does not compromise the central ethical values.

The findings of the policy recommendations (Objective 3) support policies that tackle these four main recommendations: policies on ethics and sustainability of newsroom AI application which prioritize transparency, fairness, and human regulation. It involves actual measures such as interviews with the stakeholders, workshops, and guidelines based on journalistic principles to cause minimum damage and to keep the trust of the audience. More general guidelines are based on such international rules as the Ethics of AI introduced by the UNESCO organization, which requires the formulation of methods combating inequities, insecurities, and prejudices in the media. The shift of policy to practice may be described by the example of industry, as shown by the examples of the media giants exercising their governance to earn the trust of the population. Also, other sources, such as the AI Ethics Starter Kit and the suggested guidelines on responsible AI use, offer practical tools that newsrooms can apply to resolve these problems. The connotations support the value of regulatory intervention and cooperation between the policy, press, and technologizing forces to allow AI to promote journalistic purity and safeguard it without subduing it.

It can also be seen that these discussions disclose interconnections of objectives: the workflow changes should include the ethical factor to minimize the chance to increase the biases, and the policies will serve as a scaffolding employed to make the implementation sustainable. The constraints identified on the methodology, including secondary data indicate future studies including those that are empirical and studies on the long-term impacts of AI in various cultural situations.

6 Conclusion

This chapter sheds light on the duality of AI in journalism, a strong force of innovation and a generator of deep ethical issues. AI has revolutionary possibilities to increase scalability and the involvement of the audience through the automated news production and by providing unprecedented efficiency. The policy suggestions, which have been presented herein, encourage strong ethical standards, which place human control, transparency, and inclusiveness at its core and make AI adoption reflect the primary values of journalism integrity. In AI age will be determined by a calculated trade-off one in which technology does not replace human judgment but enhances it. This synergy has the potential of creating a media that believes in truth, accountability and is service oriented to the people.

References

1. Beyond AI. (n.d.). How The Washington Post Uses Heliograf AI to Automate News Generation
2. Bodó, B., Zuiderveen Borgesius, F.J., Möller, J., Trilling, D., de Vreese, C.H., Helberger, N.: Can filter bubbles protect information freedom? Discussions of algorithmic news recommenders in Eastern Europe.Digit. Journalism **9**(9), 1130–1154
3. Braun, V., Clarke, V.: Using thematic analysis in psychology. Qual. Res. Psychol. **3**(2), 77–101 (2006)
4. Carlson, M.: The robotic reporter: automated journalism and the redefinition of labor, compositional forms, and journalistic authority. Digit. Journalism **3**(3), 416–431 (2015)
5. Carlson, M.: Automating judgment? Algorithmic judgment, news knowledge, and journalistic professionalism. New Media Soc. **20**(5), 1755–1772 (2018)
6. Chesney, B., Citron, D.: Deep fakes: a looming challenge for privacy, democracy, and national security. Calif. Law Rev. **107**(6), 1753–1819 (2019)
7. Columbia Journalism School: Tow report: "Artificial Intelligence in the News" and How AI Reshapes Journalism and the Public Arena (2024)
8. Cools, H., Diakopoulos, N., Li, C., Helberger, N., Kung, E., Zubiaga, A.: Generative AI in journalism: the evolution of newswork and ethics in a generative information ecosystem. Journalism Stud. **25**(5), 482–499 (2024)
9. Dierickx, L., Lindén, C.G., Dang Nguyen, D.T., Opdahl, A.L.: AI, Journalism, and Critical AI Literacy: Exploring Journalists' Concerns, Needs, and Responsibilities. AI & Society, Advance online publication (2025)
10. Dierickx, L., Opdahl, A.L., Khan, S.A., Lindén, C.G., Guerrero Rojas, D.C.: A data-centric approach for ethical and trustworthy AI in journalism. Ethics Inf. Technol. **26**(2), Article 15 (2024)
11. Diakopoulos, N.: Algorithmic transparency in the news media. Digit. Journalism **5**(7), 809–828 (2016)
12. Diakopoulos, N.: Transparency. In: Dubber, M., Pasquale, F., Das, S. (eds.) The Oxford Handbook of Ethics of AI, pp. 197–213. Oxford University Press (2020)
13. Fanta, A.: Putting Europe's robots on the map: automated journalism in news agencies. Reuters Inst. Chapter J. (2017)
14. Floridi, L., Cowls, J., Beltrametti, M., Chatila, R., Chazerand, P., Dignum, V., Luetge, C., Madelin, R., Pagallo, U., Rossi, F., Schafer, B., Valcke, P., Vayena, E.: AI4People—an ethical framework for a good AI society: opportunities, risks, principles, and recommendations. Mind. Mach. **28**(4), 689–707 (2018)

15. Global Investigative Journalism Network: Chaos and Credibility: A Snapshot of How AI is Impacting Press Freedom and Investigative Journalism (2025)
16. Graefe, A., Haim, M., Haarmann, B., Brosius, H.B.: Automated journalism: a meta-analysis of readers' perceptions of human-written in comparison to automated news. Media Commun. **8**(3), 50–59 (2020)
17. Guzman, A.L., Lewis, S.C.: Automation, journalism, and human–machine communication: rethinking roles and relationships of humans and machines in news. Digit. Journalism **7**(4), 409–427 (2019)
18. Helberger, N.: Towards a normative perspective on journalistic AI: embracing the messy reality of normative ideals. Digit. Journalism **11**(7), 1265–1282 (2023)
19. Helberger, N., van der Velden, L.: Designing for diversity: normative challenges in algorithmic news recommendation. Digit. Journalism **11**(7), 1283–1301 (2023)
20. IBM. (n.d.). How is AI Being Used in Journalism?
21. Kakar, S.: Digital newsroom transformation: a systematic review of artificial intelligence's influence on journalistic practices, narratives, and ethics. Journalism Media **5**(4), 1832–1852 (2024)
22. Khan, Z.: Digital transformation in journalism: mini review on the impact of AI and data-driven technologies (2014–2024). Front. Commun. **10**, Article 1535156 (2025)
23. Lewis, S.C., Westlund, O.: Big data and journalism: epistemology, expertise, economics, and ethics. Digit. Journalism **3**(3), 447–466 (2015)
24. Lindén, C.G.: Algorithms for journalism: the future of news work. J. Media Innov. **4**(1), 60–76 (2017)
25. Lindén, C.G.: A research agenda for augmented journalism. Nord. Rev. **42**(s4), 45–59 (2021)
26. Marconi, F., Siegman, A.: A day in the life of a journalist in 2027: reporting meets AI. C. Journalism Rev. (2017)
27. Media Helping Media. (n.d.). Dealing with Algorithmic Bias in News
28. Meta-Guide.com. (n.d.). Automated Journalism: Bloomberg Cyborg
29. Napoli, P.M.: What is media policy? Ann. Am. Acad. Pol. Soc. Sci. **707**(1), 27–43 (2023)
30. Nawaz, M., Hussain, S., Memon, F., Siddiqui, S.A.: Artificial intelligence in journalism: a systematic literature review of global trends, regulatory challenges, and ethical concerns. J. Digit. Media Policy **16**(1), 5–28 (2025)
31. Pariser, E.: The Filter Bubble: What the Internet is Hiding from You. Penguin Press (2011)
32. Pindrop: The Impact of Deepfakes on Journalism (2024)
33. ProfileTree: AI in News Personalisation: Tailoring Media to Individual Tastes (2025)
34. Qureshi, M.A.: AI-Driven Fact-Checking in Journalism: Enhancing Information Credibility and Reducing Misinformation. SSRN (2025)
35. Recombee: Breaking the News: The Role of AI in Modern Journalism (2023)
36. Reuters Institute: "It's a feature, not a bug"—How Journalists Can Spot and Mitigate AI Bias (2025)
37. Rini, R.: Deepfakes and the epistemic backstop. Philosophers' Impr. **20**(24), 1–16 (2020)
38. Schmelzer, R.: Beyond misinformation: the impact of AI in journalism and news. Forbes (2024)
39. Simon, F.M.: Artificial Intelligence in the News: How AI Retools, Rationalizes, and Reshapes Journalism. Tow Center for Digital Journalism, Columbia Journalism School (2024)
40. UNESCO: Ethics of Artificial Intelligence (2021)

The Use of AI in Newsrooms—A Systematic Literature Review

Hoa Vu Thi Kim

Abstract Artificial Intelligence (AI) has been reshaping journalism deeply by automating the process of collecting news, transforming. Despite being used for over a decade, AI has proved the key role in enhancing the productivity, content personalization and supported fact—checking. However, the popularity of large language models (LLMs) and automated news production system also show out the big challenges relating to ethical standards, transparency and public trust. The article has overviewed the academic works of AI applications in journalism from 2020 to 2025 with the aim of analysing 4 main apps: -Content production; -Fact checking, -Content personalization, -support edition. 51 academic materials were collected and then analyzed to clarify 3 key issues in this research: 1. The AI practical applications in news production and distribution; The operational limits and institutional challenges when integrating AI into the journalism environment. The consequence of ethics—profession due to increasing the role of algorithms in journalism. In addition, this study has found out the clear differentiation of benefits, risks and the AI approach level among the nations in Global North and Global South. The three core ethical challenges are identified in this article, including the erosion of professional independence, responsibility gap and the normalization of uncontrolled automated processes. Based on this, the research proposed an interdisciplinary ethical frame and specialized training orientation in order to integrate AI sustainably, contributing to protect the core value of journalism in the digital transformation environment worldwide.

Keywords Artificial intelligence (AI) · Journalism · News automating · Fact-checking · Content personalization · Ethic journalism · Public trust · LLMs · AI responsibility

---

H. V. T. Kim (✉)
VNU University of Social Sciences and Humanities, Hanoi, Vietnam
e-mail: Vukimhoa@vnu.edu.vn

P. Chien Thang and A. Prasanth (eds.), *AI-Assisted Journalism and Media: Opportunities and Challenges*, Studies in Computational Intelligence 1277,
https://doi.org/10.1007/978-3-032-19648-4_7

1 Introduction

1.1 Context and Rationale

The rise of AI has marked a significant turning point in all aspects of modern life, including journalism.

In 1967, Philip Meyer, a journalist at the Detroit Free Press, pioneered the use of computer-assisted data analysis to explore the causes of the Detroit riots. His innovative approach laid the groundwork for data journalism—a distinct and increasingly essential branch of modern reporting [1].

Since the early 2000s, and especially after 2010, news organizations around the world have gradually adopted AI within their production processes, covering activities from content analysis to personalized news delivery. According to the Digital News Report 2024 by the Reuters Institute for the Study of Journalism, AI has become an integral part of many newsroom operations, including automated transcription, machine translation, metadata tagging, and behavior-based content curation [2]. These technologies not only improve workflow efficiency but also enhance audience engagement and responsiveness.

A survey by Thomson Reuters Foundation [3] in 2025 shows that although more than 80% of Global South journalists use AI in daily work, fewer than 13% of newsrooms have formal policies and most lack ethical or transparency frameworks.

This situation raises serious concerns regarding the erosion of verification capacity and the perceived trustworthiness of journalistic products, particularly as algorithms operate within "black box" systems. The roles of editors and reporters are increasingly at risk of being diminished, potentially undermining investigative rigor and, more broadly, eroding journalism's professional identity and social responsibility. Furthermore, limited access to technical infrastructure and the ecosystems. The interdict culinary training may contribute to widening the gap in media ecosystems around the world.

These challenges set the urgent need to develop AI governance frameworks specifically for journalism, with the aim of protecting the transparency, professionalism, and the role of epistemic justice in media. That rapidly shifted to the digital environment.

1.2 Research Gap

Since the early twenty-first century, application of AI to journalism has grown markedly. However, lots of existing literature remains narrowly focused on AI technological dimensions, without fully analyzing ethical, professional, and social implications. Moreover, this body of research is predominantly situated within the contexts of developed countries, regions in the Global South are largely absent from global academic discourse.

There remains a lack of truly interdisciplinary and systematic synthesis that integrates three critical dimensions: (1) the practical applications of AI within newsroom workflows; (2) the ethical and professional consequences of content automation; and (3) regional disparities, particularly in postcolonial contexts and areas with constrained resources.

The study "*The use of AI in newsrooms: A systematic review of applications, limitations, and ethical challenges*" was conducted to narrow the gap by carrying out a rigorous academic review based on the PRISMA 2020 guide. The study synthesizes the popular patterns, systemized the ongoing critical debates, and proposes appropriate ethical frameworks for the future of journalism in a rapidly changing media environment to help recalibrate journalism's normative foundations in a rapidly evolving global media landscape.

1.3 The Objectives

The article critically evaluates AI's role in the field of journalism through a systematic review of academic publications and reports. Research focuses on three main parts:

- Practical AI application in news production and distribution;
- Operational and institutional limitations that constrain its implementation;
- the ethical and professional challenges emerging from the increasing algorithmic mediation of journalistic activities.

2 Research Questions

To systematically explore the intersections of AI and journalism, this study is guided by three central research questions focusing on professional practice. These three main questions are outlined in this report to examine key dimensions in journalism concluding the impact on professional practices:

- What is the AI adoption in newsrooms now, and how is it reshaping traditional journalistic workflows?
- How has the academic literature characterized the benefits and limitations of integrating AI into journalistic production?
- What ethical and professional debates have emerged in response to the AI growing influence on the news production process?

Through these questions, the study seeks not only to describe current practices but also to synthesize academic debates that explain the ethical concerns, professional dilemmas, and regional inequalities. In doing so, it aims to define a more effective, humane, and sustainable approach to AI in journalism.

3 Research Methodology

To explain the 3 above questions and to get the targets, the author used a method of systematic review to synthesize comprehensively and to analyze critically the AI-related academic works in journalism. This method enables to identify the common-applied models and to systematise the current academic discussions, also reflects knowledge gaps, especially in post-colonial areas or resource-scare regions.

The PRISMA 2020 Instruction was carried out to ensure transparency, reproducibility and scientific standards. The documents were collected from the reliable scientific databases—scopus (n = 224), web of science (n = 210) and Google scholar (n = 368)—covering the period January 2020 to December, 2025 when AI in journalism has rapidly expanded and diversifies.

Key words are used including "AI in journalism" "algorithmic journalism" "automated news generation" "robot journalism" "AI ethics in media" and they are combined by Boolean to each suitable database.

The procedure of selecting and evaluating documents consist of four steps as follows:

Identifying: A total of 802 documents were retrieved from 3 aforementioned sources.

Scanning: 673 documents were excluded due to duplication (n = 188), unrelated abstract or titles (n = 342) or insufficient academic quality (n = 143).

Evaluating the relevance: 129 documents were looked through carefully, in which 78 were removed due to not enough academic quality (n = 42), not writing in English (n = 14) or lack of report database (n = 22).

Finally, 51 scholarly selected works fulfilled the criteria: review articles or reputable industry reports, writing in English; mentioned AI directly in journalism and analyzed at least one of three main following aspects:

1. Practical profession;
2. Limits of operating and institutions;
3. Ethical and professional consequences.

The procedure was illustrated detailedly in Fig. 1.

After selecting documents, they were encoded by a standardized form to ensure their coherence. The analysis was performed using the method of thematic analysis and grouped into three main axes in accordance with research objectives:

- Practical applications of AI in journalism;
- The operational and institutional limitations;
- The issues of ethical, professional and policies

The analysis was framed within contemporary theoretical perspectives, drawing on the concept of "algorithmic power" [4] and the theory of the journalistic field [5], to examine how AI is reshaping the role of journalists. This transformation has implications for professional legitimacy and the redistribution of power within the

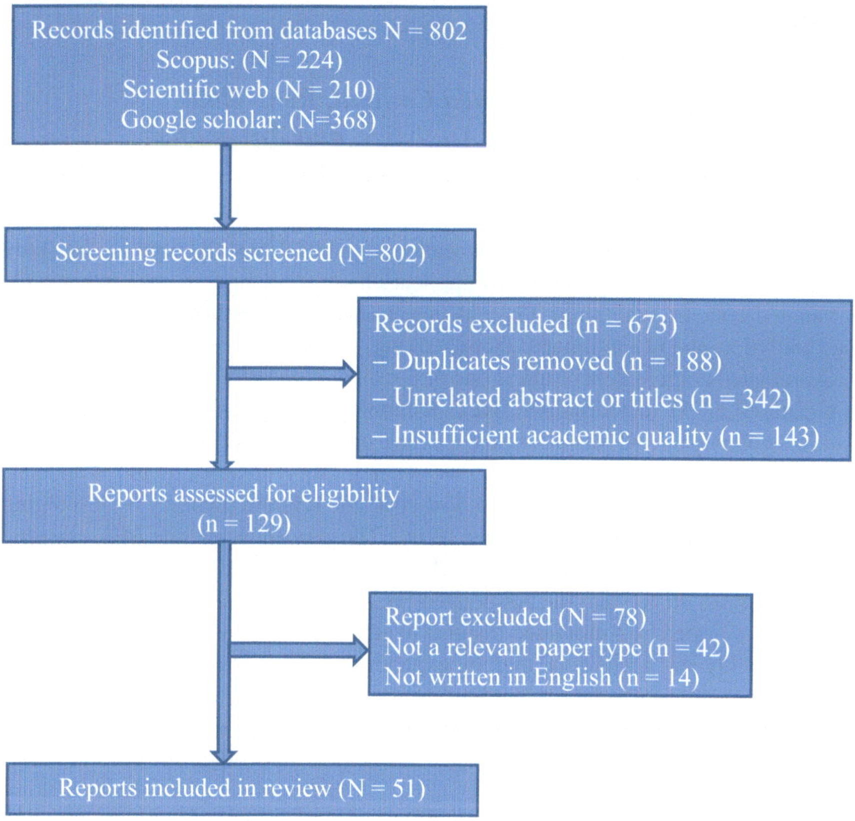

Fig. 1 The PRISMA 2020 diagram

news production process. By integrating systematic synthesis with critical theoretical reflection, this approach highlights both the academic significance and practical policy implications in a world that is rapidly transitioning toward technology-driven journalism models.

Despite not studying deeply on the comparison of the typical cases, the process of synthesis explained clearly about the significant background differences in *Global North* and *Global South*. Such factors as technological infrastructure, specialization level, political-social conditions become an outstanding variable which controls AI integration efficiently and sustainably. In this situation, "Algorithmic power" not only reflects the change of reporting tools but also is the symbol of shifting the professional authority structure of modern journalism.

4 Results

4.1 Applications: AI-Generated Content, Real-Time Transcription, Audience Analytics, and Automated Fact-Checking

Based on the synthesis of 51 selected academic publications and reports, the study identified four main categories of AI applications in contemporary newsrooms: (1) automated content generation, (2) real-time transcription and translation, (3) user data analytics, and (4) automated fact-checking.

Automated content generation is one of the most pioneering and clearly effective applications of AI in journalism. NLG technologies such as those developed by Automated Insights and Narrative Science have enabled newsrooms to automatically produce thousands of short-form news stories (e.g., sports results, financial updates, weather information) with high speed, low cost, and reliable accuracy by leveraging structured data inputs (e.g., scores, data feeds). These systems operate 24/7, ensuring that news content is generated immediately at a large scale [6]. Wang and Dodds [7], *also* cited the evidence that the Associated Press (USA) has utilized the Wordsmith platform developed by Automated Insights to generate thousands of regular financial reports quickly and accurately, thereby expanding its capacity for producing short-form content. Similarly, The Guardian Australia used the Reporter Mate system during federal elections to automatically create localized content based on public datasets, ensuring both timely publication and tailored coverage for each electoral district [7].

Another study by Friday and Soroaye [8] reports that The Washington Post used Heliograf during the 2016 Rio Olympics to produce real-time news updates with integrated anomaly detection features, supporting the newsroom in delivering fast and accurate reporting during major events. In Spain, 20minutos and La Vanguardia have maximized automated news production about traffic accidents using data from emergency services. However, these reports are typically unverified, lack diverse perspectives, and are often direct reproductions of press releases from police or fire departments [9].

CH Media (Switzerland) has used AI to create short news of local voting results or regional-level sports ones [10]. These news stories are published with little or no manual editing, mainly to quickly inform local audiences. Every day, these systems can produce hundreds of articles, reducing the role of human reporters [11].

In 2024, La-Rosa and Sandoval-Martín conducted a Turing experiment which demonstrated that AI-generated news content was rated as equally reliable as that produced by professional journalists, although it was found to be less vivid and engaging compared to human-written content [12].

In the news production process, newsrooms have employed integrated voice-based artificial intelligence systems, in concluding speech recognition, translation, and voice synthesis technologies. These systems can automatically record speeches and interviews, convert written news into audio bulletins, support visually impaired users or those requiring hands-free access, and generate podcasts from news articles [13]. This technology is particularly useful in real-time scenarios such as interviews, press conferences, and live-streamed videos. According to a global survey conducted by the Thomson Reuters Foundation [14], more than 80% of journalists reported using AI tools like ASR (Automatic Speech Recognition) to assist with note-taking, transcription, and content editing.

However, despite its high application potential, Voice AI systems still face certain technical limitations: AI-generated voices (Text-to-Speech TTS) may lack accuracy, especially when dealing with regional accents or domain-specific terminology; AI voices also induce lower levels of brain activity compared to human speech, resulting in reduced listener attention and memory retention. This negatively impacts communication effectiveness and reduces audience trust in the news Additionally, research by Kuhn et al. [15] highlights those factors such as background noise, number of speakers, and language diversity can significantly affect the Word Error Rate (WER). These findings showed the need for human oversight in practical situations.

However, in countries of the Global South, the adoption of AI in newsroom production remains relatively limited. Jamil [16] observed that AI implementation in Pakistan is still in its early stages—despite its potential, progress is slowed by a lack of awareness, insufficient investment, and professional skepticism. Many journalists still do not fully understand what AI is or how it can be practically applied.

Moreover, numerous news organizations lack the infrastructure, software, and financial resources necessary to integrate AI technologies. Concerns also persist among journalists that AI could replace human roles or undermine the quality of content.

At present, there are no comprehensive training programs or clear national strategies for incorporating AI into journalism. According to Trang et al. [17], only a few media organizations in Vietnam have adopted AI in a systematic or strategic manner. The use of AI in Vietnamese newsrooms remains inconsistent, lacks clear direction, and often depends heavily on individual journalists' personal initiative.

In addition, Vietnamese journalists apply AI technology very little. They use AI for only grammar checking, translating and content summarizing.

AI expresses strategic potential in audience analytics by processing, analyzing user behaviors to personalize content and optimize news transmission. A study clarified that 62% newsrooms used AI to personalize news content by Sonni et al., by using user data and behavior insights, AI enables personalized experiences through audience separation and prediction model, showing an increasing reliance on audiences basing one algorithm [18].

AI assists editorial teams in reaching and analyzing their audiences by identifying interests, habits, timing, behaviors, and suitable types of content [19]. The article also incorporates psychological theories and reception models to explain how audiences accept AI-driven personalization in journalism.

Brannon et al. [20] proposed an "audience view"—an AI-powered platform based on large language models (LLMs) designed to aggregate audience feedback on a large scale. This system automatically identifies key topics, connects comments to

4.3 Ethics in Integrating AI into Digital Journalism: Three Core Challenges

AI-integration into content production in digital newsrooms was widely discussed in recent research. 90% percent of the surveyed studies highlighted the ethical challenges for news organizations. These studies identify three critical ethical concerns about: the loss of professional autonomy, the responsibility gap, and the normalization of uncontrolled automation. These challenges not only affect journalistic practice but also raise profound questions about morality, power, and transparency in the modern media landscape.

The first challenge is the loss of autonomy. AI's ability to automatically suggest, prioritize, and content selection has diminished the journalist's active editorial role. While AI can generate highly convincing texts, it often lacks transparency in how information is created, offers no clear way to verify the accuracy of input data, and is prone to reproducing biases embedded in its training data. This leads to generating false or misleading content, presenting information without sufficient context for accurate interpretation. As a result, AI-generated outputs may decrease the public trust in content produced by news organizations.

Using AI in newsrooms is likely to change journalists' roles—from being in full control to mostly just supervising. Instead of deciding what messages to share, they may be influenced or even directed by algorithms [33].

The accountability gap is a key issue raised in many studies as many newsrooms lack clear mechanisms to verify AI-generated content. AI often works as a system that spreads or blurs responsibility making it unclear who is accountable: the algorithm developers, the editors, or the technical managers. This breaks down transparency, which is a core value in journalism. AI has become a kind of "responsibility-avoidance device" [34], weakening public trust and making it harder to hold anyone accountable. This creates a serious ethical and legal loophole in the system.

The increasing automation of newsroom operations, when introduced without strong oversight or systematic critical evaluation, has resulted in what scholars term the "naturalization of technology" [35]. Under these circumstances, AI systems are rarely questioned or scrutinized, eliminating vital opportunities for self-correction, error detection, and meaningful editorial discussion. Gradually, an emotionless, machine-driven approach to content creation may become normalized, functioning with little or no human intervention. From a media sociology standpoint, this trend cultivates a mindset of "non-resistance" among journalists and audiences alike, diminishing critical thinking and undermining professional ethics [35]. These subtle yet damaging consequences call for structural interventions and collaboration among diverse stakeholders.

The interdependence of these challenges presents a compounding ethical risk. As Prunkl [36] observes, threats to human autonomy rarely occur in isolation but tend to interact and mutually reinforce one another. In automated news production, the erosion of human autonomy can directly exacerbate the accountability gap, while insufficient oversight of automation further intensifies problems. This interplay underscores the urgency of establishing a cross-disciplinary ethical framework in which journalists, developers, and policymakers collaborate to safeguard human agency and restore editorial control in an increasingly automated media environment.

5 Discussion

5.1 Theoretical Contribution: Algorithmic Power and a Professional Structure in Journalism

The above findings confirm that AI is reshaping power structures in processing news and distribution. Based on the theoretical framework of algorithmic power proposed by Couldry and Hepp [4], the increasing use of machine learning systems and language models moved authority from traditional form to automated system platforms.

In journalism, professional authority is traditionally rooted from various forms of cultural, symbolic, ethical and social recognition and ability to shape public discourse [5]. However, because AI has become embedded in main newsroom functions such as news production, fact-checking, and audience analytics, the important traditional forms of capital has declined, challenging journalism's established authority. As Møller et al. [37] argue, AI not only disrupts routine tasks but also reshapes the definition of professional authority, requiring the profession to reinforce core ideals in algorithmic contexts, readjust educational curricula, and reclaim specialized work practices to sustain its legitimacy in the evolving media landscape.

5.2 Three Ethical Challenges and their Mutual Relationships

The findings also highlight three serious ethical challenges: loss of professional autonomy, accountability gap, and the normalization of unchecked automation. These phenomena are not isolated, they are deeply interconnected.

The loss of control contributes to blurred the boundaries of responsibility, while unconditional automation reinforces the inevitability of the other two challenges. As Evans et al. [38] have warned, when AI is deeply embedded in social processes without clear ethical oversight or accountability mechanisms, it becomes a "responsibility-dispersing device," creating legal and moral uncertainty.

In such environments, journalists have turned from being producers of information to supervisors. This raises a critical question: Can key journalistic values such as truth, transparency, and public accountability still be kept when publishing by AI systems driven by performance optimization logic rather than ethical standards?

5.3 Conflict Between Technological Advancement and Ethical Integrity in Journalism

The rise of Al poses a critical dilemma for journalism: whether to embrace new technologies to remain relevant or to safeguard professional standards in order to preserve public trust. Research shows that while many newsrooms have gained remarkable speed and efficiency through Al adoption, this progress has also increased their reliance on algorithms as "filters of reality." Since Al systems are trained on historical data, they may inadvertently reproduce biases or steer content toward particular narratives. Consequently, core journalistic principles-such as accuracy, objectivity, and fairness-are increasingly at risk.

The development of AI in newsrooms is going to be at risk of turning the communication process into a chain of audience behavior optimization, rather than serving the public benefit. This runs counter to the core principles of journalism as a democratic institution.

5.4 Challenges in the Global South

The findings from this systematic review the highlight of a deeper structural layer of digital power in global journalism: the underdevelopment of newsrooms in the Global South stems not only from limitations in infrastructure or human resources, but also from the cumulative effects of a global technological architecture, one that is designed, operated and standardized according to the conditions and interests of the Global North. This serves as a clear illustration of the phenomenon of "technological colonialism", in which Global South countries are not only confined to the role of technology recipients but are also excluded from shaping how these technologies are integrated into journalism practice.

In developing countries, the unclear rule and tightly controlled media make it harder to use Al in journalism. The slow use of Al tools for fact-checking, content monitoring, and automation reflects not only technical limitations but also the lack of clear management. Furthermore, many journalists in the Global South do not know much about AI, so they depend on tools made by foreign technology companies. This reduces local control and gives more power to global technology platforms in setting the standards for journalism.

From a theoretical view, the study shows how AI acts as a structural force that is reshaping journalism—changing its core practices, shifting power and redefining professional relationships. By combining the idea of "algorithmic power" [4] with a postcolonial perspective, the article develops an interdisciplinary framework that explains both how AI is present and why its influence is uneven across the global media environment.

Therefore, evaluating the impact of AI in journalism requires more than just looking at efficiency or automation. The experience of Global South countries shows that adopting technology must be understood within local contexts, in which history, politics and institutional strength shape the outcomes.

5.5 Implications for Policy and Education, Moving Toward an Interdisciplinary Ethical Model

To mitigate these risks, newsrooms need to establish an interdisciplinary ethical framework that engages three main groups: journalism, system designers, and policy-makers. This framework should be gilded by key requirements; responsibility should be clearly defined at every stage of news productions and journalism training should include AI literacy covering bias detection, system auditing, and critical engagement with data.

In education, training institutions should embrace a clear philosophy: in future, journalists are not only narrators but also "data auditors" who need to combine insights from sociology, ethics, data science to maintain professional autonomy in a media environment shaped by algorithms.

Building on the above findings, it is clear that AI is driving a deep structure shift within newsrooms. Emerging AI technologies are reorganizing news production, pushing them toward greater data, personalized content, and advanced automation.

In future, AI is going to function as an "assistive editorial intelligence", helping with real-time content choices, identifying emerging topics through social trend analysis, and tailoring news feeds to specific audiences. Newsrooms may also create "proactive updates" that use social sensing to anticipate public information demands before they are explicitly searched for. This development would advance a journalism model focused on delivering personalized, real-time, and milti-format content spanning text, audio, and visual media.

The growing integration of AI with extended reality (XR) and immersive journalism introduces new chances for developing news experiences that respond to both emotions and behaviors. AI technologies can tailor content and adjust its delivery to different reception contexts ranging from mobile devices to smart speakers and virtual platforms, thereby increasing relevance and audience engagement across multiple formats.

To remain resilient in the technology era, newsrooms must design and implement AI systems that align with principles of social responsibility and transparency. Future newsrooms should also establish positions such as "data editors" or "ethics coordinators", tasked with monitoring AI applications, evaluating algorithmic bias, protecting professional standards and upholding openness across the entire news production workflow.

Rather than viewing the relationship between humans and machines as antagonistics. Future journalism should advance toward a model of responsible human—AI collaboration. This model seeks to align AI's capabilities with ethical values, critical judgement and professional independence of journalists.

5.6 Theoretical and Practical Contributions, and Research Limitations

This article shows ongoing scholarly debates by extending Couldry and Hepp's [4] theoretical framework of algorithmic power through the incorporation of Bourdieu's [5] notion of field specific capital in journalism and interdisciplinary ethical perspective. By merging social field theory with ethical analysis, the study illustrates that AI should be understood not simply as a technological instrument but as a structural force that reshapes power dynamics, redefines professional legitimacy, and alters the processes of symbolic capital formation in contemporary news production.

Practically, the study recommends establishing roles like "data editors" and "ethics coordinators" to manage AI systems, enhance transparency, mitigate bias, and protect journalism's core values. Additionally, the study suggests a new educational philosophy for training journalists, highlighting the necessity of integrating knowledge from sociology, ethics, and data science in order to maintain professional agency in an increasingly algorithm-driven media environment.

However, this study has several limitations. The analysis primarily relies on a synthesis of academic literature and case reports, without qualitative data from actual newsrooms or in-depth interviews with relevant stakeholders.

Therefore, future research should delve into the lived experiences of journalists, system developers, and policymakers to validate and expand upon the conclusions presented in this article.

5.7 Research Limitations and Future Directions

Although this study has provided a systematic and theoretical ground of AI role in journalism, several limitations must be acknowledged to ensure academic integrity and to pave the way for future research.

First, publication bias is a notable limitation. The majority of academic literature analyzed in this study focuses on successful or promising cases of AI adoption. Failed practices, technical obstacles, or organizational conflicts are often underreported, resulting in a lack of critical perspectives and counter-evidence in the current discourse.

Second, the study is constrained by the scope of database access. Although the academic sources like Scopus, Web of Science, and Google Scholar were extensively utilized, grey literature including internal newsroom reports, unpublished policy documents, and research findings outside traditional academic systems was largely excluded. These sources may offer valuable content in actual implementation as well as AI operational dynamics within newsrooms.

Third, language limitation is a noteworthy constraint. The study primarily draws upon English-language sources, which may limit access to local perspectives especially in non-English-speaking nations where the application or perception of AI might differ significantly due to distinct cultural, political, and media contexts.

To address the aforementioned limitations, future research can pursue three main directions. First, conducting cross-national comparative studies particularly between Global North and Global South countries can help identify contextual elements which shape and form AI adoption in journalism. Second, combining qualitative and quantitative methods such as in-depth interviews with journalists, developers, and policymakers alongside large-scale surveys can enhance and validate current findings. Third, case studies of internal AI policies in newsrooms, especially in developing are built, implemented, and adjusted in real-world conditions.

6 Conclusion: Academic Contributions and Practical Implications

This study not only offers a comprehensive mapping of AI current adoptions in journalism, but also contributes to a systemic overview of the interplay with technology, ethics, power in an algorithmically mediated media environment. By integrating the theoretical frameworks of algorithmic power and capital structure within the journalistic field, the paper sheds light on how AI is reshaping professional authority, redistributing power in the news production process, and posing unprecedented ethical challenges to journalism integrity.

The significance of this study lies not only in its analysis of the current landscape but also in its proactive proposal of an interdisciplinary ethical framework and a strategic approach to journalism education tailored to the data-driven era. This framework emphasizes the incorporation of transparency, accountability, and auditability into the design and operation of AI systems. At the same time, it envisions a new generation of journalists—professionals equipped with data-critical thinking and technological oversight skills to navigate and shape the evolving media ecosystem responsibly.

More importantly, this paper raises a systemic policy question: How can Global South countries avoid becoming "passive consumers" in the AI-driven transformation of journalism and instead shape their own technological sovereignty and development strategies? In response, the study offers a clear roadmap for capacity-building in infrastructure, ethical policy frameworks, and training programs especially in resource-constrained and legally underdeveloped contexts such as Pakistan, Vietnam, and other developing nations.

With its interdisciplinary approach, deep theoretical grounding, and clear practical orientation, this study contributes not only to academic discourse on journalism and technology, but also offers actionable policy insights for shaping a human-centered, transparent, and accountable journalism ecosystem in the age of AI.

Declaration of Generative AI Use During the preparation of this work, the author used Elicit, NotebookLM, and ChatGPT (OpenAI, GPT-5 version) to support literature organization and language refinement. The author reviewed and verified all content and take full responsibility for the final version of the manuscript.

References

1. Meyer, P.: Precision Journalism: A Reporter's Introduction to Social Science Methods, 4th edn. Rowman & Littlefield Publishers, Lanham (2002)
2. Newman, N., Fletcher, R., Robertson, C.T., Eddy, K., Nielsen, R.K.: Digital News Report 2024. Reuters Institute for the Study of Journalism. https://reutersinstitute.politics.ox.ac.uk/digital-news-report/2024 (2024)
3. Thomson Reuters Foundation: Journalism in the AI era: opportunities and Challenges in the Global South and Emerging Economies. TRF Insights, Jan 2025. https://www.trust.org/wp-content/uploads/2025/01/TRF-Insights-Journalism-in-the-AI-Era.pdf (2025)
4. Couldry, N., Hepp, A.: The Mediated Construction of Reality: Society, Culture, Mediatization. Polity Press, Cambridge (2017)
5. Bourdieu, P.: On Television and Journalism. Pluto Press, London (1998)
6. Schapals, A.K., Porlezza, C.: Assistance or resistance? Evaluating the intersection of automated journalism and journalistic role conceptions. Media Commun. **8**(3), 16–26 (2020). https://doi.org/10.17645/mac.v8/3.3054
7. Wang, S., Dodds, D.: Harmonizing traditional journalistic values with emerging AI technologies: a systematic review of journalists' perception. Media Commun. **13**, Article 9495 (2024) https://doi.org/10.17645/mac.9495
8. Friday, J.P., Soroaye, M.P.: The Rise of AI Journalism: How Algorithms are Shaping News Content (2024)
9. Olivar-Julián, F.J.: Automated journalism in news about traffic accidents: emergency services information dump. Texto Livre. **17**, e49369 (2024). https://doi.org/10.1590/1983-3652.2024.49369
10. Aargauer Zeitung: Wie wir mit Datenanalyse und Automatisierung über den Regionalfussball berichten. Retrieved from https://www.aargauerzeitung.ch/sport/kurzberichte-wie-wir-mit-datenanalyse-und-automatisierung-den-regionalfussball-abdecken-ld.2177851 (27 Aug 2021)
11. Beck, C.: So arbeiten Roboter bei den Medien mit. Persoenlich.com. Retrieved from https://www.persoenlich.com/medien/so-arbeiten-roboter-bei-den-medien-mit (22 Jan 2023)
12. La-Rosa Barrolleta, L.A., Sandoval-Martín, T.: Artificial intelligence versus journalists: The quality of automated news and bias by authorship using a Turing test. Analisi: Quaderns de Comunicación Cultura **70**, 15–36 (2024) https://doi.org/10.5565/rev/analisi.3681
13. Simon, F.M.: AI in the newsroom: lessons from the adoption of the globe and mail's Sophi. J. Pract. (2025). https://doi.org/10.1080/17512786.2025.2471781
14. Thomson Reuters Foundation: Journalism in the AI Era: Opportunities and Challenges in the Global South and Emerging Economies. Thomson Reuters Foundation, London. Available at https://www.trust.org/wp-content/uploads/2025/01/TRF-Insights-Journalism-in-the-A-Era.pdf (2025). Accessed 7 Aug 2025
15. Kuhn, K., Gerber, R., Widmer, M., Zaugg, T.: Measuring the accuracy of automatic speech recognition solutions. ACM Trans. Access. Comput. **16**(4), Article 25 (2024) https://doi.org/10.1145/3636513
16. Jamil, S.: Artificial intelligence and journalistic practice: the crossroads of obstacles and opportunities for the Pakistani journalists. J. Pract. **15**(1), 1–23 (2020). https://doi.org/10.1080/17512786.2020.1788412

17. Ta, T.N., Pham, C.T., Le, D.H., Vi, T.P., Tran, Q.Q.: Understanding the adoption of artificial intelligence in journalism: an empirical study in Vietnam. SAGE Open. **14**(2), 21582440241255241 (2024). https://doi.org/10.1177/21582440241255241
18. Sonni, A.F., Hafied, H., Irwanto, I., Latuheru, R.: Digital newsroom transformation: a systematic review of the impact of artificial intelligence on journalistic practices, news narratives, and ethical challenges. J. Media. **5**(4), 1554–1570 (2024). https://doi.org/10.3390/journalmedia5040097
19. Lim, J.S., Zhang, J.: Adoption of AI-driven personalization in digital news platforms: an integrative model of technology acceptance and perceived contingency. Technol. Soc. **69**, 101965 (2022). https://doi.org/10.1016/j.techsoc.2022.101965
20. Brannon, W., Beeferman, D., Jiang, H., Heyward, A., Roy, D: AudienceView: AI-assisted interpretation of audience feedback in journalism. In: Companion of the 2024 Conference on Computer-Supported Cooperative Work and Social Computing (CSCW '24 Companion) (2024) https://doi.org/10.1145/3678884.3681821
21. Beckett, C., Yaseen, M.: Generating Change: A Global Survey of What News Organisations are Doing with Artificial Intelligence. JournalismAI (London School of Economics) (2023)
22. Quelle, D., Bovet, A.: The perils and promises of fact-checking with large language models. Front. Artif. Intell. **7** (2024). https://doi.org/10.3389/frai.2024.1341697
23. García-Méndez, S., de Arriba-Pérez, F., Barros-Vila, A., González-Castaño, F.J., Costa-Montenegro, E.: Automatic detection of relevant information, predictions and forecasts in financial news through topic modelling with latent Dirichlet allocation. Appl. Intell. **53**(16), 19610–19628 (2023). https://doi.org/10.1007/s10489-023-04452-4
24. Amponsah, P.N., Atianashie, A.M.: Navigating the new frontier: A Comprehensive Review of AI in Journalism. Adv. J. Commun. **12**, 1–17 (2024)
25. Peng, X., Xu, Q., Feng, Z., Zhao, H., Tan, L., Zhou, Y., Zhang, Z., Gong, C., Zheng, Y.: Automatic news generation and fact-checking system based on language processing. arXiv:2405.10492 [cs.CL]. https://doi.org/10.48550/arXiv.2405.10492 (2024)
26. Jones, B., Jones, R., Luger, E.: AI 'everywhere and nowhere': addressing the AI intelligibility problem in public service journalism. Digit. J. **10**(10), 1731–1755 (2022). https://doi.org/10.1080/21670811.2022.2145328
27. Xavier, B.: Biases within AI: challenging the illusion of neutrality. AI Soc. **40**, 1545–1546 (2024). https://doi.org/10.1007/s00146-024-01985-1
28. Ntoutsi, E., Fafalios, P., Gadiraju, U., Iosifidis, V., Nejdl, W., Vidal, M.-E., et al.: Bias in data-driven artificial intelligence systems—an introductory survey. Wiley Interdiscip. Rev. Data Min. Knowl. Discov. **10**(3), e1356 (2020). https://doi.org/10.1002/widm.1356
29. Grimme, M., Zabel, C.: AI in the newsroom: a collective case study about newsworker-AI collaboration in the German newspaper industry. J. Media Bus. Stud. (2024). https://doi.org/10.1080/16522354.2024.2380120
30. Adjin-Tettey, T.D., Muringa, T.P., Danso, S., Zondi, S.: The role of artificial intelligence in contemporary journalism practice in two African countries. Journalism Media. **5**(3), 846–860 (2024). https://doi.org/10.3390/journalmedia5030054
31. Soto-Sanfiel, M.T., Ibiti, A., Machadom, M., Marín Ochoa, B.E., Mendoza Michilot, M., Rosell Arce, C.G., Angulo-Brunet, A.: In search of the global south: assessing attitudes of Latin American journalists to artificial intelligence in journalism. J. Stud. **23**(13), 1649–1668 (2022). https://doi.org/10.1080/1461670X.2022.2088308
32. Savage, N.: Fact-finding Mission: artificial intelligence provides automatic fact-checking and fake news detection, but with limits. Commun. ACM. **64**(3), 18–19 (2021). https://doi.org/10.1145/3446879
33. Noain-Sánchez, A.: Addressing the impact of artificial intelligence on journalism: the perception of experts, journalists and academics. Commun. Soc. **35**(3), 105–121 (2022). https://doi.org/10.15581/003.35.3.105-121
34. Cheong, B.C.: Transparency and accountability in AI systems: safeguarding wellbeing in the age of algorithmic decision making. Front. Hum. Dyn. **6**, 1421273 (2024). https://doi.org/10.3389/fhumd.2024.1421273

35. Gutiérrez-Caneda, B., Lindén, C.-G., Vázquez-Herrero, J.: Ethics and journalistic challenges in the age of artificial intelligence: talking with professionals and experts. Front. Commun. (2024). https://doi.org/10.3389/fcomm.2024.1465178
36. Prunkl, C.: Human autonomy at risk? An analysis of the challenges from AI. Minds Mach. (2024). https://doi.org/10.1007/s11023-024-09665-1
37. Møller, L.A., Skovsgaard, M., de Vreese, C.: Reinforce, readjust, reclaim: how artificial intelligence impacts journalism's professional claim. Journalism. (2024). https://doi.org/10.1177/14648849241269300
38. Evans, K.D., Robbins, S.A., Bryson, J.J.: Do we collaborate with what we design? Top. Cogn. Sci. **15**(5), 879–898 (2023). https://doi.org/10.1111/tops.12682

Navigating Ethical Boundaries: The Use of AI in Journalistic Practice

Nguyen Thi Thanh Huyen, Bui Vinh Khang, and Dao Thanh Hang

Abstract As a natural cause of the increase in adopting technology and tools in journalistic practice, specifically artificial intelligence (AI), a series of challenges appear such as algorithmic bias and disinformation, putting journalism under the risk of eroding its core values. Accuracy and transparency considered as journalistic ideals and ethics are needed to be strictly ensured. In this chapter, by integrated methods such as document analysis and in-depth interview, the current state of AI usage and ethics is examined in some Southeast Asian countries. As indicated, journalists and reporters in this region use AI at most of the news and content production stages, such as research, data analysis, fact-checking, and grammar or spelling editing. However, the adoption is now lacking the regulations at both national and organisational levels. With the combination of theoretical and practical approach, this study provides not only the description but also recommendations for policy and professional practice in order to construct an ethical, transparent and professional journalism landscape in Southeast Asia.

Keywords AI ethics · Digital journalism ethics · Southeast Asia

T. T. H. Nguyen
School of Journalism and Communication, University of Social Sciences and Humanities, Vietnam National University, Hanoi, Vietnam

V. K. Bui (✉)
Communications Consultant, United Nations Population Fund, Hanoi, Vietnam
e-mail: vinhkhang.bui01@gmail.com

T. H. Dao
Journalism and Communication Management, School of Journalism and Communication, University of Social Sciences and Humanities, Vietnam National University, Hanoi, Vietnam

P. Chien Thang and A. Prasanth (eds.), *AI-Assisted Journalism and Media: Opportunities and Challenges*, Studies in Computational Intelligence 1277,
https://doi.org/10.1007/978-3-032-19648-4_8

1 Introduction

AI is now becoming a popular tool used in many areas of life. Journalism is undoubtedly inevitable in that flow. With this technology, the news has been changed in the way it is produced and distributed. Through a series of prompts, our journalists are supported in tasks such as producing materials or tailoring content, especially when using the large language models (LLMs) such as ChatGPT. AI can even identify the upcoming trends by analysing massive datasets. A recent survey found that most major news organizations worldwide are now using AI in their daily operations [1, 2].

Everything has pros and cons, and AI is no exception. It offers efficiency and speed, but also many complex questions related to ethics. A question is globally raised to answer: Who is responsible for the inaccuracies of AI-generated news—journalists or AI itself? Besides, the reliance on AI may cause the erosion of critical thinking and fact-checking skills, which true journalists have to develop everyday to protect their expertise. Such questions are being topics to discuss in the conferences for media and journalism practitioners, in Western countries, and in countries with diversity of culture and ethics systems such as Southeast Asia.

2 History of the Research Problem: AI Ethics in Journalism

Scholarly interest in the relationship between AI and journalism has been growing since the mid-2010s. However, the conversation shifted dramatically with the advent of large language models (LLMs). The public release of GPT-3 in 2020 and even more so the launch of ChatGPT in late 2022 sparked an urgent, global debate over the complex ethical questions arising at the intersection of AI and journalism.

2.1 Global Research Dynamics and Trends

International research on AI in journalism has followed three main paths, moving from a focus on technological potential to a deeper look at ethical challenges.

The first path explored how AI could reshape journalism. Early on, Carlson (2015) introduced the idea of algorithmic journalism [3], and Graefe (2016) showed that AI could be used across the entire news process from gathering data and writing stories to editing and distributing them [4]. Even in those early years, though, questions of transparency and accountability began to surface.

The second research trajectory focused on the project's hazards and ethical challenges. Diakopoulos (2019) cautioned about algorithmic bias and how AI could promote social prejudices [5]. More recently, UNESCO (2023) stated that journalists are cautiously adopting Generative AI. As for quality news outlets, accuracy and credibility are paramount. Therefore, worries about the "hallucinations" of Generative AI outputs are top of mind. Human engagement and monitoring are necessary when using Generative AI to avoid false or misleading information from being spread. In response, organizations like Reuters Institute (2024) have urged the creation of ethical guidelines to protect the credibility of information [2].

Finally, the third research path looks at how journalists and newsrooms are adapting. Although over 80% of journalists we surveyed are using AI, the majority were self-taught and only 13% reported that their newsrooms had an AI policy in place [6]. According to sharing in the webinar hosted by UNESCO, AI is beginning to transform journalism in South-East Asia by streamlining tasks like transcription and translation, and enhancing investigative work. However, its rapid, unregulated adoption is raising urgent concerns about bias, misinformation, and the skills journalists need to navigate this new landscape [7]. With over 60% of Vietnamese media agencies having integrated AI tools in 2024, however, most applications currently focus on content creation functions such as headline suggestions, editing, translation and multimedia generation [8]. This reveals a critical policy vacuum that needs to be filled urgently as news organizations struggle to catch up with the pace of technological change.

2.2 Regional Context and Research Gaps in Vietnam and some South East Asian Countries

In Southeast Asia, AI in journalism is still in its early stages, largely focused on experimentation and adapting to new technology. While international reports such as Journalism, Media, and Technology Trends and Predictions 2024 (Reuters Institute, 2024) recognize the region's push for digital change, many ethical problems have yet to be solved [9]. Issues such as a lack of transparency, the spread of misinformation, and declining public trust are widely recognized as critical challenges.

In Malaysia, studies by Sarji and Abdul Aziz (2025) and Sualman (2025) document the use of AI in data analysis, fact-checking, and editing support. However, these researchers emphasize the urgent need for clear ethical guidelines, enhanced training on AI tools, and stronger commitments to transparency [10, 11]. The widespread use of AI, if not carefully managed, risks further undermining the public's confidence in news.

In Thailand, Hongladarom (2021) examined the country's National AI Ethics Guidelines as a sign of public policy modernization, revealing tensions between technological globalization and traditional values [12]. Fieldwork by Wongrujira (2025) indicates that ChatGPT is used only to a limited extent in newsrooms, mostly for ideation or light content generation [13]. Yet reliability issues, lack of proper sourcing, and persistent ethical concerns remain key barriers. The study recommends investment in data infrastructure, capacity-building for journalists, and the development of context-specific codes of conduct for AI use in journalism.

In Indonesia, Besman and Evita (2024) argue that while AI enhances productivity, it cannot replace core journalistic values such as objectivity and content quality [14]. They emphasize the need for a proactive strategy to integrate AI into journalism, with public trust placed at the center. Nevertheless, human oversight and cross-checking mechanisms remain essential to ensure ethical standards and content integrity.

In Vietnam, academic research on AI in journalism remains limited. Although some newsrooms have begun experimenting with AI tools, there is no official evaluation of their effectiveness or ethical risks. Despite the government's issuance of a National AI Strategy through 2030, there are still no specific guidelines for the media sector [15]. The current Code of Ethics for Journalists by Vietnam Journalists Association does not address new ethical dilemmas posed by AI [16]. As a result, many journalists find themselves "feeling their way forward," navigating the use of technologies such as data-driven investigations, automated quote extraction, or content generation without a clear ethical framework.

Overall, research in the region has so far focused on describing the current landscape and warning of potential risks. In-depth analyses of AI governance, reader protection policies, or sustainable business models for AI-assisted content remain scarce. While international scholarship has established a solid theoretical foundation regarding the intersection of AI, journalism, and ethics, Southeast Asia is still in the process of catching up. This highlights the urgent need for more systematic, comparative research across the region aimed at developing ethical frameworks tailored to local political and cultural contexts.

3 Understanding of Morality and Professional Ethics

As AI becomes increasingly integrated into all areas of life-especially journalism-it is crucial to examine the ethical dimensions of technological application. Ethics in this context must be understood not only as universal principles, but also through the lens of professional responsibility, reflecting the journalist's role in society.

From an etymological perspective, the word "moral" (Latin: moralis) is rooted in mos/mores, meaning customs or social norms. It emphasizes the communal and practical aspects of right and wrong, in contrast to "ethics," which tends to align with more abstract philosophical theories. From Aristotle's Virtue Ethics to Kant's Deontology and the framework of Consequentialism, philosophers have long upheld ethics as a guiding compass for responsible, transparent, and fair action [17, 18].

In Southeast Asia, moral reasoning is deeply influenced by cultural and religious traditions. Buddhism, for instance, promotes Right Livelihood and Right Speech as part of the Noble Eightfold Path-values that align with journalistic principles such as honesty, objectivity, and commitment to the public good [19, 20]. Confucianism emphasizes the Five Constants (Ngũ Thường), particularly Righteousness (Yi), Wisdom (Zhi), and Trustworthiness (Xin)-core attributes of integrity, critical thinking, and reliability in journalism (Fung, 1948; Ivanhoe and Van Norden, 2001)

[21, 22]. Islamic thought also contributes to journalistic ethics through principles such as justice (‘adl), truth (haqq), and communal responsibility (amanah) [23, 24].

Building on these moral foundations, professional ethics in journalism are generally grounded in four key principles: truthfulness, fairness, transparency, and public accountability. In the AI era, these responsibilities become even more urgent. Journalists must disclose when content is AI-generated, avoid using AI to generate deceptive or harmful information, and deliberately avoid algorithmic bias caused by incorrect or unbalanced training data. Finally, the journalist is responsible for the content, not the AI. Ethics is a supportive tool guiding people to values. You are journalists, a reporter, or an individual belonging to society, you have to follow those high ideals and principles. Especially, you have the social responsibility of delivering the truth and promoting a trustworthy information environment. When approaching and using new technology such as AI in your work, it is necessary to ethically use, even if you serve as a journalist purely for the purpose of earning money, professional ethics is the key to help you survive over job security.

4 Findings and Discussion

4.1 Current Practices of AI Integration in Southeast Asian Newsrooms

The debate on using AI responsibly currently becomes a wider dialogue across Southeast Asian countries. At what level should journalists rely on AI? Should audiences be allowed to know which article is generated by AI? And who is accountable when AI produces inaccurate or misleading information? These are some of the hot topics being discussed in many regional forums nowadays.

Both journalists and the public have expressed growing concern over the accuracy and trustworthiness of AI-generated content, two pillars central to journalistic integrity. Moreover, large language models (LLMs), trained on vast datasets, may inadvertently reproduce bias or misinformation, further eroding public trust in the media.

Southeast Asia. In February 2024, ASEAN released the Guide on AI Governance and Ethics, a set of recommendations aimed at businesses across the region. Its primary goal is “to promote alignment within ASEAN and foster interoperability of AI frameworks across jurisdictions.” The document emphasizes a voluntary and flexible approach, drawing extensively on examples from Singapore. However, it remains unclear whether other member states will adopt and implement the guide in a consistent manner [25].

A survey conducted by Chatrine Siswoyo and Raphael Lachkar among media professionals in ASEAN highlighted three major areas of concern and and reveals concerns about the use of AI described Fig. 1 [26]:

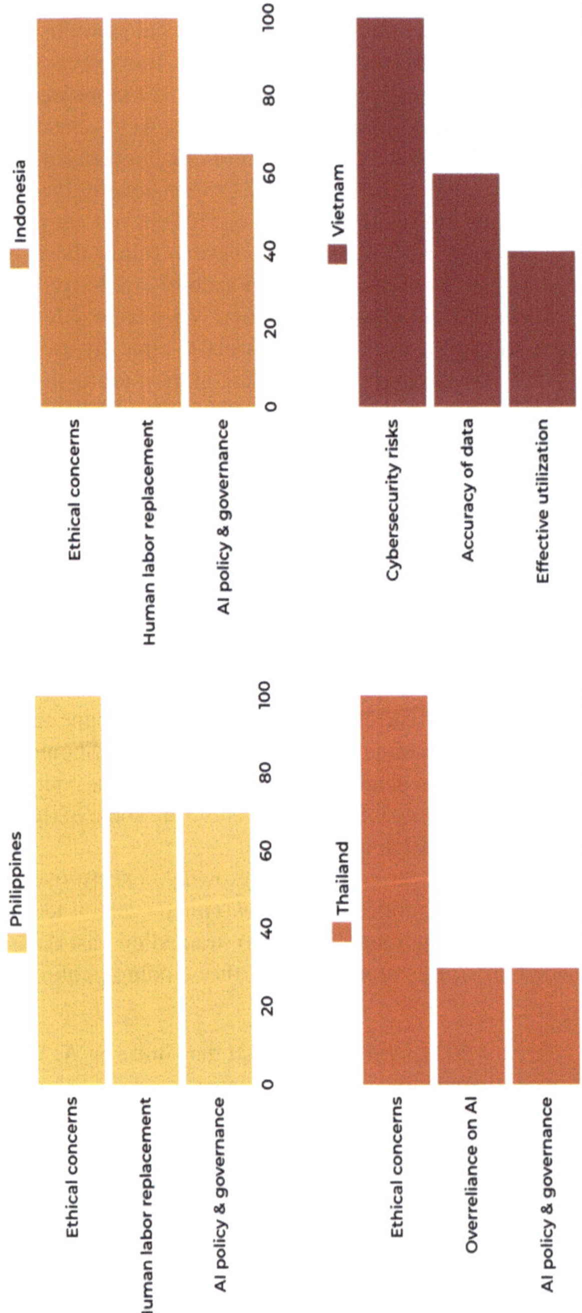

Fig. 1 Top 3 concerns about the use of AI

The risk of labour displacement and erosion of human skills: Many journalists fear that AI could overshadow the role of human agency, leading to the decline of core competencies such as investigative reporting, fact-checking, and critical analysis-skills essential to upholding journalistic integrity.

Issues of quality, accuracy, and copyright: AI-generated content may contain factual inaccuracies or infringe upon intellectual property rights, especially when models are trained on unlicensed datasets. The examples related to the voice of Scarlett Johansson and OpenAI highlight the urgent need for regulatory frameworks.

Algorithmic bias, misinformation, and a lack of transparency: AI can amplify biases in training data, leading to skewed or misleading results. Furthermore, ambiguity about the role of AI in content development blurs the line between human-produced and machine-generated news.

Based on this sense, it is critical for stakeholders to collaborate in order to ensure the responsible integration of AI. Newsrooms must establish open and well-defined editorial processes for the use of AI, and press associations should give ethical guidelines that account for regional peculiarities. Meanwhile, regulators must aggressively implement supervision mechanisms to defend the public interest (Table 1).

The question has shifted from "What can AI do?" to "What should AI do?" and "How can we govern it to serve society best while preserving the integrity of journalism?" Journalists, parliamentarians, technological businesses, and members of the general public should all participate in a multi-stakeholder dialogue. The goal is to use technology to support journalism, not to undermine its fundamental ideas.

Indonesia. Newsrooms in Indonesia are utilizing AI in a way that strikes a balance between creativity and a careful, moral approach. While they're leveraging AI to boost efficiency and make content more interactive, domestic media companies are also setting up frameworks to ensure the technology is used responsibly.

For example, since 2018, IDN Media has used "Timmy the Robot," an AI journalist, to automatically write short news stories. This system turns raw data like weather reports, earthquake updates, or sports scores into simple, easy-to-read articles. Launched in December 2023, "Intinya Sih," a hugely successful initiative, employs AI to distill long content into concise summaries. This meets the needs of younger Indonesian viewers who demand concise social media information. As

Table 1 Overview of AI regulatory frameworks in selected Southeast Asian countries [27]

	AI law enacted	Draft AI law/regs	AI guidelines	National AI strategy
Cambodia	No	No	No	No
Indonesia	No	No	No	Yes
Malaysia	No	No	No	Yes
Philippines	No	Yes	No	Yes
Singapore	No	No	Yes	Yes
Thailand	No	Yes	Yes	Yes
Vietnam	No	Yes	No	Yes

seen Fig. 2 [28], the initiative's outcomes were outstanding, demonstrating a 12.5% increase in engagement and an incredible 244.8% growth in advertising impressions in a few months.

However, there are considerable concerns regarding the ethical difficulties that accompany this achievement. In June 2025, the Indonesian Press Council proactively established guidelines on the use of artificial intelligence in journalism. These standards, developed after extensive consultation with media outlets and specialists, are intended to uphold the key journalistic principles of independence, justice, and accuracy while also ensuring openness and moral behavior in the use of AI. The requirement for human assessment of AI-generated journalistic work is a significant component of the rules established by the prominent Alliance of Independent Journalists (AJI). The AJI has stated that AI should only be used as a supplement and cannot wholly replace journalism made by humans in Indonesia [29].

Despite the popularity of initiatives such as "Timmy the Robot," IDN Media regulates Timmy's output using a separate AI system developed by a different business and applying unique training models. Furthermore, IDN Media's editorial, product, technical, and audience development teams oversee this implementation, acting as a system of checks and balances.

Cambodia. AI applications are empowering Cambodian newsrooms to deliver information in both Khmer and English with unprecedented speed and clarity. This effectively breaks down language barriers and expands their audience reach. A prime example is the launch of Cambodia's first AI-powered Khmer news application, which garnered over 35,000 queries in just 24 hours [30]. In contrast, when interviewing a Laos journalist, AI tools are not popular in most newsrooms due to the fact that there is no AI tool supporting Laos language, therefore, only reporters from English-written newspapers use AI for grammar checking in their daily work. Cambodian journalists are leveraging AI to automate translations, create subtitles, and summarize content, helping critical stories reach a broader readership.

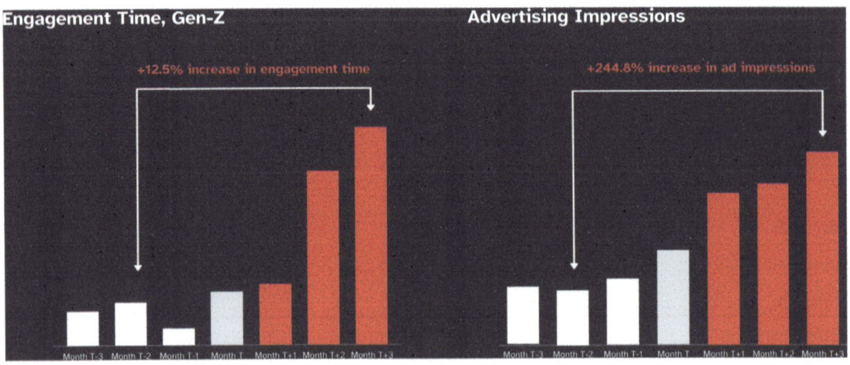

Fig. 2 Engagement time and advertising impressions before/after implementation of the "Intinya Sih" AI initiative on IDN Times [28]

However, the integration of AI into Cambodia's media ecosystem also presents significant ethical challenges. Concerns primarily focus on the risk of AI-generated fake news, deepfakes, and manipulated images, particularly in sensitive areas such as politics and public safety. The Ministry of Information and media oversight bodies are emphasizing the need for transparent AI practices, human oversight for sensitive news, political reporting, and investigative journalism, and adherence to ethical codes, especially regarding AI use, to maintain public trust and prevent misinformation [31].

Cambodia is actively working to develop its first national AI strategy, aiming to balance innovation with the protection of journalistic freedom, truth, and cultural heritage. To boost credibility and fight misinformation, new ethical guidelines and codes of conduct are being put in place. While AI offers immense potential for improving efficiency and creativity, Cambodian experts unanimously agree that AI cannot replace human judgment and critical thinking in investigative journalism or in-depth storytelling.

Thailand. Thai media is changing in a big way, and AI is a major force shaping the future of journalism. Alongside traditional challenges like fake news and the rise of content creators on social media platforms, AI is perceived as one of the three biggest threats to Thai journalism in 2025, according to experts at the Foreign Correspondents' Club of Thailand (FCCT) [32]. However, instead of avoiding it, media organizations here are proactively building strategies to adapt and leverage AI. Their goal is to both enhance competitiveness and strengthen transparency and public trust.

The application of AI in news production processes is becoming increasingly common. According to Ms. Pensopa Sukontarak, a senior reporter at the Thai Journalists Association, advancements in Natural Language Processing (NLP) and machine learning have significantly boosted the development of AI journalism. Newsrooms are using AI to assist in automating routine tasks such as data collection and verification, information storage, text editing support, and the production of illustrative graphics. AI is also seen as a tool to increase efficiency and accuracy, reducing processing time for repetitive tasks like large data analysis or interview transcription. This allows journalists to focus more on in-depth analysis and investigation [33].

A prime example is Sanook.com, Thailand's most visited news site with over 40 million monthly users. Sanook has adopted AI transparently to maintain its leading position for over two decades. The newsroom believes the crucial issue isn't whether to use AI, but how to use and optimize this technology responsibly [32]. Another example is Thairath TV, the first organization to use the MiMir system for news image archiving. This system can automatically tag, transcribe, translate, and recognize faces, significantly speeding up the process of searching and utilizing archival material. Concurrently, Thairath has also developed a personalized content recommendation system based on AI, which has increased its Click-Through Rate (CTR) from 1.5 to 4%. AI technology is also leveraged in analyzing viewer behavior to optimize digital advertising [33].

However, the application of AI in Thai journalism also raises numerous ethical questions. Mr. Chavarong Limpattamapani, Chairman of the National Press Council, warns that AI needs to be operated with control mechanisms, especially in information verification and disclosing AI's role in content production. He emphasizes that AI should be a "powerful assistant, not a replacement for humans" [33]. The risk of AI unintentionally amplifying fake news due to reliance on unverified input data is a current challenge. This requires newsrooms to maintain human oversight and commit to transparency at all stages of the AI-powered journalistic process.

Vietnam. In recent years, AI has unlocked unprecedented opportunities for journalism in Vietnam, driving digital transformation and enhancing content production quality. At the 2025 National Press Forum, many experts expressed optimism about AI's potential to help newsrooms innovate their workflows and engage audiences more effectively in the digital media environment. According to a survey by Nguyen Quang Dong (Institute for Policy and Development of Journalism), the proportion of Vietnamese news agencies that have applied or plan to apply AI nearly doubled between 2023 and 2024, reaching over 60% [34].

However, the actual level of application remains modest, primarily focusing on basic stages of content production. Popular platforms such as ChatGPT, Gemini, or Copilot are mainly used for suggesting headlines, drafting content, summarizing, grammatical editing, creating illustrative images or videos, and assisting with translation. Most organizations utilize free or low-cost tools, with limited financial investment and technological infrastructure [34].

Despite this, some large news agencies have implemented AI relatively comprehensively. Kinh te and Do thi Newspaper integrates AI into almost all stages of its news production process, from suggesting topics, summarizing and editing, to analyzing user behavior and automating repetitive tasks [35]. VietnamPlus (VNA) has been using the Wochit platform since 2016 for video production, and since 2021, it has expanded strategic cooperation to enhance the effective use of AI in content personalization and automated news bulletins [36]. Lao Dong Newspaper has developed a "virtual editor" system capable of translating content into English and generating subtitles for videos [36]. VnExpress has also applied AI in content personalization, summarizing news in text and voice formats, and using robots to compile financial and sports news [36]. Notably, Vietnam Economic Times/VnEconomy has developed an AI ecosystem called the Marcom-AI Platform and established the D.A Alliance to integrate AI into its content management system (CMS), with specialized tools such as an economic chatbot, an automatic summarizer, an economic context translation tool, and a text-to-speech tool [37]. This model is designed to control input data quality and extend applications beyond journalism to include businesses and public services.

In terms of policy, Vietnam has not yet enacted a comprehensive legal framework for AI. Although the National Strategy on Research, Development, and Application of AI by 2030 (Decision No. 127/QD-TTg) has set goals for establishing research centers and supportive institutions, the related legal system is still in its formative stages [27]. Current regulations are mainly integrated from existing documents on intellectual property, cybersecurity, and personal data protection. Recently, the Ministry of Information and Communications announced a Draft National Standard on the AI Lifecycle to promote transparent and responsible applications [27]. While there is an intention to develop a legal framework, the government's current approach prioritizes economic and technological objectives, with a tendency to limit regulations that might hinder AI development. As Yen Vu (2024) noted, "There is currently no comprehensive AI legal framework in Vietnam, but the country has a clear intention to develop one in the near future. In the meantime, existing regulations on intellectual property, IT, e-commerce, cybersecurity, and data protection may be applied temporarily." [38].

Despite its potential, Vietnamese journalism still faces three key challenges when it comes to AI. First, AI development isn't yet a top priority. Most news outlets are focused on using AI for basic content creation rather than exploring how it could help them understand reader behavior or develop new business models. Second, the adoption of AI is fragmented and disorganized. It often happens spontaneously within different organizations rather than as part of a coherent, long-term strategy. Finally, many newsrooms simply lack the necessary resources such as money, staff, and technical infrastructure to sustainably integrate AI into their operations.

Because of this, most of the current research and applications in Vietnam just describe the situation and identify ethical risks. We still need more in-depth studies on things like AI governance, user protection policies, and the real economic benefits of AI-powered content. To build a truly responsible AI journalism ecosystem in Vietnam, we'll need a long-term vision and serious investment in institutional frameworks, technology, and human talent.

4.2 Perceived Ethical Dilemmas and Challenges

As AI becomes more common in journalism, journalists across Southeast Asia face many ethical and practical challenges. Journalists interviewed in Vietnam, Thailand, Laos, and Cambodia reveal a wide range of opinions regarding AI. Some journalists are optimistic, some are neutral, and others are deeply concerned.

Most see AI as a helpful tool that can boost productivity, handle repetitive tasks, and process data quickly. However, they all agree that AI can't replace humans when it comes to critical thinking, emotional intelligence, or professional ethics. A few journalists remain distrustful or concerned because they don't fully understand how the technology works.

When it comes to learning about AI ethics, these journalists get their information from a mix of sources, including internal training, workshops from press associations, and academic materials. However, the effectiveness of widespread training courses remains limited because "the needs of each newsroom also differ" and the reality that "the average age of experienced reporters is very high, leading to limited effectiveness in absorbing new knowledge" (according to a Hanoi based journalist).

One of the foremost ethical issues is accuracy and the risk of misinformation. Journalists are well aware that if input data is not verified, AI can generate false or misleading information. Specific cases have occurred; for example, a journalist used AI to create a virtual character without verification, leading to serious consequences for credibility.

Furthermore, authorship and transparency in AI use are contentious issues. A journalist from Tien Phong (Vietnam) believes that "we certainly must state that this article used AI," while a journalist from Voice of Vietnam radio argues that "AI should not be granted authorship, because AI is just a tool, lacking creative consciousness or legal responsibility like humans." Therefore, authorship should belong to the programmer or the end user, the person who directly uses AI to create content, i.e., the journalist, editor, or anyone who inputs prompts, selects data, and generates the output product using AI.

Privacy and data security are also major concerns. Journalists worry that sensitive information could be stored or leaked by AI. Although there is awareness of encrypting or anonymizing sources, there is currently a lack of clear national regulations on protecting input data when using AI in journalism.

Another challenge is over reliance on AI, which can diminish personal skills and critical thinking. When conducting the indepth interview, our group faced an issue that a journalist had asked for the list of interview questions, then sent to us the AI-generated answers. From that reality, it's not just suggesting headlines or outlining structures. At the higher risk, it will impact on the distinct writing style. When mentioning this example, we do not call for stopping AI in content production, we advocate for fostering inventive approaches and new forms of narrative through using AI responsibly and ethically.

The challenge is more complex when evaluating the performance of journalists in the newsroom and compensating for their work. Looking into the equity in labor, it becomes unfair when paying for journalists totally using AI and journalists writing the news by themselves at the same rate. It may lead to workplace conflict.

There are also differences among newsrooms in adopting AI tools. Some newsrooms allow a limited number of tools, even do not allow their journalists to use AI for creating content. In contrast, others ask their reporter to use AI for producing podcasts, editing video, and consider AI as a strong assistive tool in quality and error checking. However, AI is not broadly applied, just at individual level. Therefore, the common understanding in AI is important, aimed at fostering AI usage in newsrooms, then collaboratively developing specific ethical policies and encouraging journalists to apply technology in their daily work.

5 Conclusion

AI is increasingly integrated into journalism not only worldwide but also in Southeast Asia. Beside its advantages and opportunities, AI also causes ethical concerns that require the strong administration from authorities and the deep understanding of journalism from practitioners and policymakers.

It is easily seen that most journalists use AI for routine tasks, such as seeking ideas, searching, checking grammar and spelling, even some use for generating content. However, there are differences among journalists and countries in this issue. As mentioned, senior journalists with a long time of working are often reluctant to adopt new tools, compared to the eagerness of young ones. Some newsrooms allow and encourage their journalists to use AI, others strictly restrict AI for content generation. In countries like Indonesia and Cambodia, there are domestic AI products supporting the work of journalists, meanwhile no AI tool supports Laos language. Limited investment and reliance on free software have resulted in a lack of a clear strategy.

Across the region, journalists may get lost in the complexity of raising concerns. Whether they accept the reality that AI is a helpful tool and need adapting in daily work, or they have to neglect this tool due to being scared of losing their skills such as fact-checking, critical thinking, and on-site reporting. In any situation, human responsibilities are top priority and AI can not replace that role.

Lazy thinking and writing is becoming a big concern within journalism, especially as many journalists nowadays get familiar with AI tools and become over-reliant, leading to the erosion in skills. AI-generated articles or videos can result in misinformation, which is regarded as a big ethical mistake. Labeling the journalistic work should be taken under consideration to ensure the transparency and intellectual property attribution. AI tools are still continuous under the process of learning, therefore, their data can cause algorithmic bias. Furthermore, AI poses a risk at data privacy and security when journalists use it to analyse sensitive information.

The technique is not wrong, but doing without ethics is. To protect the mission of journalism, a clear ethical framework and a strong governance system is necessary to be discussed, however, it is fragmented among Southeast Asian countries. Basically, the code of ethics is widely used and inclusive, therefore, adapting the framework in each country needs to ensure the suitability for unique cultural and socioeconomic position in each country.

6 Recommendations

6.1 Policy Recommendations

Building a credible media landscape and the age of AI requires collective action from regulators, news organizations, and training institutions. These stakeholders must design and implement updated professional standards. Such measures aim to balance the advantages of AI applications with journalism's enduring principles of accuracy, openness, and public accountability.

The first step toward responsible AI in journalism is to build a strong legal and ethical foundation. National guidelines created through collaboration with press organizations can be modeled on international principles such as the Paris Charter or UNESCO framework. They must be tailored to the specific context of each country. These rules should clarify that publishers remain accountable and that humans retain ultimate control AI over processes. Every instance of AI generated or heavily AI assisted output should be transparently labeled.

Instead of waiting for regulators, newsrooms should take the lead in creating their own rules for AI. These internal policies might include approved tools, clear boundaries for usage, and explicit bans. In particular, Nhan Dan Newspaper restricts AI in content creation, meanwhile Hanoi Moi adopts a more open approach toward new technologies. With expenditures on licensed AI tools having the function of controlling every sub-accounts, organizations can track the utilisation of each individual such as reviewing the prompts or identifying the level of using the tool. By managing sub-accounts, both leaders and journalists are well informed about performance evaluation, since then the tracking report helps to address and resolve ethical issues. In advance, newsrooms should build strong review systems that encourage cooperation between editorial, technical, and product teams. When managing sensitive political or social reporting, this mission becomes more important. Continuous feedback from both staff and the politic will keep these policies relevant. Moreover, ethical impact assessments carried out before AI is introduced can identify risks early and outline mitigation strategies.

Finally, the collaboration among stakeholders in establishing the AI application standard in journalism are essential. Regulators, newsrooms, journalistic groups, technological experts, and academics must have a shared voice in forums advocating policy in journalism, ensuring that the core values are retained through ethical practice. The ethical implications of AI in journalism with an emphasis on local settings and the risk monitoring model are necessary to be researched aimed at providing a solid scientific framework for policy development and practical improvements.

6.2 Professional Practice Recommendations

As Francis Bacon, "knowledge is power". Every good thing is rooted in education. To begin, newsrooms and universities must provide a fundamental understanding of AI, including definition, process, and impact, to both working and future journalists. That approach is straightforward to defend a profession from ethical misconduct. Along with basic knowledge, journalists should be trained across many skills such as analysing the AI-generated content, critical thinking, style of writing and using the prompts effectively. Within each organisation, AI application standards and best practices must be addressed, taught and shared, since then media leaders could encourage the prolonged learning and ethical AI usage.

Secondly, the process of creativity and news production need to be under control in order to ensure human oversight is always present. It is highlighted that AI is a tool, and a tool can not replace human role. The responsibility of journalists is examining before publication so as not to cause misinformation. This prevents the journalist from overreliance on AI.

Regarding policymakers, it's essential to establish an appropriate legal framework to regulate AI use in journalism, including provisions on algorithmic transparency, data privacy, and legal liability in cases of misinformation.

Thirdly, increasing transparency and accountability is essential for building public trust. When using AI at any stage, from translation and data processing to spell checking, journalists must clearly indicate the tools used and the extent of AI's intervention. Despite AI's assistance, the ultimate responsibility still rests with the journalist. As has been pointed out: "When a reporter makes a mistake, they are accountable. But when AI causes an error, responsibility often dissolves into the becoming anonymous organization" [39]. This highlights that the journalist's role in oversight and prudence becomes even more critical in the digital age. Furthermore, extreme caution is needed when providing sensitive information to AI systems, especially when clear internal regulations are lacking.

Media development organizations and donors can contribute by investing in locally relevant AI training programs, supporting both professional journalism and freelance journalists, and promoting pilot projects on ethical AI applications in news production.

Finally, fostering critical thinking and adaptability across the entire team is paramount. Instead of fearing AI, journalists should proactively approach it as a supportive tool. Professional performance evaluation should not be based on the extent of AI use, but rather on the quality, accuracy, creativity, and social impact of journalistic work. In the future, we might consider: should we train AI according to journalistic ethical principles to make AI an "ethical assistant" that reflects the values journalists strive for?

In addition, technology platforms and developers also bear the responsibility of designing AI tools that meet journalistic needs, not just focusing on content generation capabilities, but also supporting verification, ensuring fairness, and respecting local languages and cultures. Transparency in how AI produces results and sharing training data are also essential for building trust from journalists and the public.

Declaration of Generative Ai and Ai-Assisted Technologies in the Chapter Preparation Process
During the preparation of this work, the author(s) used Gemini in order to perform grammar and spelling checks, and to ensure the accuracy and appropriateness of industry-specific terminology. After using this tool/service, the author(s) reviewed and edited the content as needed and take(s) full responsibility for the content of the published article.

References

1. Newman, N., Fletcher, R., Schulz, A., Andı, S., & Nielsen, R. K.: Reuters Institute Digital News Report 2023. Reuters Institute for the Study of Journalism. (2023)
2. Newman, N., Fletcher, R., Schulz, A., Andı, S., & Nielsen, R. K.: Reuters Institute Digital News Report 2024. Reuters Institute for the Study of Journalism. https://www.reutersinstitute.politics.ox.ac.uk/digital-news-report/2024 (2024)
3. Carlson, M.: The robotic reporter: automated journalism and the redefinition of labor, compositional forms, and journalistic authority. Digit. journalism **3**(3), 416–431 (2015). https://doi.org/10.1080/21670811.2014.976412
4. Graefe, A.: Guide to automated journalism. Tow Center for Digital Journalism, Columbia University. https://www.cjr.org/tow_center_reports/guide_to_automated_journalism.php (2016)
5. Diakopoulos, N.: Automating the News: how Algorithms Are Rewriting the Media. Harvard University Press (2019)
6. Larbi, L.: World Press Freedom Day: the need for the equitable and ethical adoption of AI. Thomson Reuters Foundation. https://www.trust.org/2025/05/02/world-press-freedom-day-the-need-for-the-equitable-and-ethical-adoption-of-ai/ *Thomson Reuters Foundation* (2025)
7. James, M.: 'AI cannot be curious. It has no heart': UNESCO Bangkok convenes journalists to debate the future of the newsroom. UNESCO. https://www.unesco.org/en/articles/ai-cannot-be-curious-it-has-no-heart-unesco-bangkok-convenes-journalists-debate-future-newsroom (2025)
8. Dharmaraj, S.: Vietnam: Balancing AI Innovation and Ethics in Journalism. OpenGOV. https://archive.opengovasia.com/2025/06/21/vietnam-balancing-ai-innovation-and-ethics-in-journalism/ (2025)
9. Newman, N.: Journalism, media, and technology trends and predictions 2024. Reuters Institute. https://reutersinstitute.politics.ox.ac.uk/journalism-media-and-technology-trends-and-predictions-2024 (2024)
10. Sarji, N.A., Abdul Aziz, A.: The usefulness of artificial intelligence (AI) on the efficiency of news delivery towards journalists in Klang Valley. J. Media Inf. Warf. **18**(1), 28–40 (2025)
11. Sualman, I.: Artificial intelligence (AI) technology and its relevance to the profession and ethics journalism in Malaysia. Jurnal Kewartawanan Malaysian Press Institute, 1(1), 130 (2025)
12. Hongladarom, S.: The Thailand national AI ethics guideline: an analysis. J. Inf. Commun. Ethics Soc. **19**(4), 480–491 (2021)
13. Wongrujira, M.: A survey on journalists' use and perceived impact of ChatGPT on Thai newsrooms. Communication and Media in Asia Pacific (CAMP), 8(1), 1–16 (2025)
14. Besman, A., Evita, N.: AI in semi-automated journalism: a review of Indonesia's journalistic ethics. J. Commun. Stud. **8**(3), 583–594 (2024)
15. Thủ tướng Chính phủ: Quyết định số 1668/QĐ-TTg về Chiến lược quốc gia về nghiên cứu, phát triển và ứng dụng trí tuệ nhân tạo đến năm 2030. Văn bản pháp luật Việt Nam. https://vanban.chinhphu.vn/portal/page/portal/chinhphu/hethongvanban?class_id=1&_page=1&mode=detail&document_id=210461 (2021) [Government of Vietnam: Decision No. 1668/QD-TTg on the National Strategy for Research, Development, and Application of Artificial Intelligence by 2030 (2021)]
16. Hội Nhà báo Việt Nam: Bộ Quy tắc đạo đức nghề nghiệp người làm báo Việt Nam. https://hoinhabaovietnam.vn/bo-quy-tac-dao-duc-nghe-nghiep-nguoi-lam-bao-viet-nam/ (2016). [Vietnamese Journalists Association Code of Professional Ethics forVietnamese Journalists: (2016)]
17. Driver, J.: Consequentialism. Routledge (2012)
18. O'Neill, O.: Constructions of reason: explorations of Kant's practical philosophy. Cambridge University Press (1989)
19. De Silva, P.: Exploring Buddhist Ethics. Oxford University Press (2005)
20. Tonni, S.T.N.: Buddhist Ethics and its Impacts on Modern Time. University of Rajshahi (2021)

21. Fung, Y.-L.: A Short History of Chinese Philosophy. The Macmillan Company (1950)
22. Ivanhoe, P.J., Van Norden, B.W. (eds.): Readings in Classical Chinese Philosophy. Seven Bridges Press (2000)
23. Zaheril, Z.: Islamic new media ethics. Online J. Commun. Media Technol. **7**(1) https://www.ojcmt.net (2017)
24. Fakhry, M.F.: Ethical theories in Islam. Brill (1991)
25. Tan, A., Wright, B.: The evolution of AI regulation in Asia: a comparative analysis. Elastic Blog. https://www.elastic.co/blog/artificial-intelligence-regulation-asia-comparative-analysis#asean. (2024)
26. Lachkar, R., Siswoyo, C.: AI and journalism in Southeast Asia: a survey of opportunities and challenges. Vero Asean.. https://vero-asean.com/whitepaper/ai-and-journalism-in-southeast-asia-survey/. (2024)
27. Harriman, J.: AI policy and regulatory frameworks take shape in APAC. BowerGroupAsia. https://bowergroupasia.com/ai-policy-and-regulatory-frameworks-take-shape-in-apac/. (2023)
28. Lau, J.-N.: Indonesia's IDN media leveraging AI to offer greater value, user experience. WAN-IFRA.. https://wan-ifra.org/2024/05/indonesias-idn-media-leveraging-ai-to-offer-greater-value-user-experience/. (2024)
29. Bagus, R., & Kenzu: Indonesia sets standards for AI use in journalistic works. Antara. https://en.antaranews.com/news/342706/indonesia-sets-standards-for-ai-use-in-journalistic-works (2025)
30. Times, K.: Cambodia's first Khmer language AI news app logs 35,000+ user queries within 24 h of launch. https://www.khmertimeskh.com/501696394/cambodias-first-khmer-language-ai-news-app-logs-35000-user-queries-within-24-hours-of-launch/#google_vignette (2025)
31. Raksmey, H.: Information ministry: Press freedom improving, AI represents challenges. Phnom Penh Post https://www.phnompenhpost.com/national/information-ministry-press-freedom-improving-ai-represents-challenges (2025)
32. Leishman, R.: 'New strategies' needed for Thai media. Bangkok Post. https://www.bangkokpost.com/thailand/general/2972723/new-strategies-needed-for-thai-media (2025)
33. Toumi, H.: What relations between journalism and AI? Thailand examples stressed at WJC 2024. The Asian. https://theasian.asia/archives/183657 (2024)
34. Vietnam Plus: Vietnamese newsrooms urged to adopt strategic AI integration amid digital shift. https://en.vietnamplus.vn/vietnamese-newsrooms-urged-to-adopt-strategic-ai-integration-amid-digital-shift-post321367.vnp (2025)
35. Nguyen, N.: Redefining journalism: how AI is powering Vietnam's newsrooms. HanoiTimes. https://hanoitimes.vn/redefining-journalism-how-ai-is-powering-vietnam-s-newsrooms.704681.html (2025)
36. Nhan Dan Newspaper. Journalism in the age of AI. Vietnam.vn. https://www.vietnam.vn/en/bao-chi-thoi-ai (2025)
37. Thuong, H.: A step towards AI-powered journalism. VNEconomy. https://vneconomy.vn/a-step-towards-ai-powered-journalism.htm (2025)
38. Vu, Y., Thammasujarit, P., & Baranda, E.: What's happening in AI-related legislation across Asia? Part 2. Rouse. https://rouse.com/insights/news/2024/what-s-happening-in-ai-related-legislation-across-asia-part-2 (2024)
39. The Guardian a robot wrote this entire article. Are you scared yet, human? https://www.theguardian.com (2021)

Building AI-Ready Newsrooms

Ramya Anandanatarajan, R. Saktheeswari, and C. Yaashuwanth

Abstract In today's world most of the fields including engineering, healthcare, transportation, finance, marketing, cybersecurity has been influenced by Artificial Intelligence (AI). AI has made a tremendous changes and wonders. Domain of journalism is no exception. Journalism has also been greatly influenced and improvised by AI. Some of the uses of AI in newsroom include automatic content generation, AI-powered storytelling, Data Analysis, Speech to Text conversion, fake news deduction and content personalization. Even though AI has improvised the standard of journalism there are few concerns like data integrity, misinterpretation, employees upskilling and liability. Currently some of the media industry lack financial support, AI literacy, and specialized support from technicians. Some of the other issues are algorithm openness, ethical challenges, data privacy and accountability. In future most of the countries will depend on Artificial Intelligence for news delivery. Most of the nations have considered integrating Artificial Intelligence into their newsrooms in a positive sense, while others are still considering adoption of Artificial Intelligence. Adopting AI means, the newly designed system should be able to apply machine learning or deep learning, natural language processing and delivery without any language barrier. A complete knowledge of generative AI and Large Language Model is not known to most of the users. It can cause troubles. In order to solve this potential issue, regulatory bodies should define rules, create awareness and provide necessary guidelines to make sure true information is being shared to the viewers/users. This is a sign of healthy integration of AI and journalism.

Keywords Artificial Intelligence · Journalism · Newsroom 3.0 · Machine Learning · Media

---

R. Anandanatarajan (✉)
Department of Electronics and Communication Engineering, Sri Venkateswara College of Engineering, Chennai, India
e-mail: ramyaa@svce.ac.in

R. Saktheeswari · C. Yaashuwanth
Department of Information Technology, Sri Venkateswara College of Engineering, Chennai, India

P. Chien Thang and A. Prasanth (eds.), *AI-Assisted Journalism and Media: Opportunities and Challenges*, Studies in Computational Intelligence 1277,
https://doi.org/10.1007/978-3-032-19648-4_9

1 Introduction

Traditional media newsroom is known as Newsroom 1.0 which was pre-digital, that is print and broadcast techniques were dominating the market. Then came the era of Newsroom 2.0 (beginning of twenty-first century) in which digital media like online news, social media came into existence. The audience had freedom to express their likes and dislikes, views immediately. This gave a way for the media to track the expectations or nature of the audience.

Present day media are naturally designed for mobile users, social media and interactive technologies, where a current technology gives them a key point in the news ecosystem. Traditional media and digital media differ in their methods and formats. While traditional media solely depends on experienced professionals and long story boards, digital media focuses on simple, short, multimedia content customized for smart devices and social attractions. Editorial sections are moving swiftly to prioritize speed of publishing thereby pushing the journalist to perform a multifunctional role from collecting, consolidating, researching, and editing to publishing with additional role online in future. The boundaries of traditional publishers are to be expanded to include technology advancements in delivering valuable content. Now we are moving towards Newsroom 3.0 which is a complete game changer. It is going to integrate Artificial Intelligence (AI), Machine Learning (ML), Natural Language Processing (NLP), automation and many more such features. With these features it is possible to have tailored news content and multiplatform distribution with better speed and efficiency.

Success now depends on the adoption of promising technologies in designing advanced newsrooms for engaging users and revenue generating models. Few organizations have considered adopting models tailored to linking performance based on compensation models which drives earnings through user engagement and retention. Eventually staying active means accepting innovation, without compromising core journalism principles and ethics. To excel in the upcoming years, media enterprise has to grow using latest technology and at the same time it has to ensure its credibility. Today's media is growing technology-wise because of implementation of Artificial Intelligence (AI), Machine Learning (ML), Natural Language Processing (NLP), big data, data analytics, content recommendations, augmented reality, virtual reality, blockchain and cyber security. AI is an essential component and it automates the whole process in journalism. This chapter of the book deals with the various capabilities and functionalities of AI in designing the future news delivery pattern. It also investigates how AI is contributing in content creation and its editorial procedure. On the other hand, it also delves into the hurdles created due to AI adoption in terms of accuracy, ethics and technical difficulties. It shows a way to apply AI in journalism with accountability and transparency. This chapter gives a clear idea on the latest news generation and delivery techniques addressing the needs of academicians, researchers, and stakeholders.

1.1 Problem with Digital News Delivery System

In present digital era, the overflowing amount of news globally generated every second across various domains has ended up in a state of information dump. Readers are pushed with a flood of unwanted updates which fills their mail boxes or chats and drags them into mental state where they disengage themselves from any news sources especially mobile devices with limited screen size and attention span. A Reuters report in 2023 revealed that more than 30% of global population admitted that overflowing unwanted news occasionally or periodically let them to avoid news due to its overhype and negative emotional impact. Lack of personalized filtering option further aggravates the problem. Major streaming platforms are flooding the users with unwanted updates. Giants like Facebook, twitter X tends to customize the services provided through content customization but it is algorithm driven with the goal of generating more views, likes, shares and comments which aims to increase the user engagement. As a result, users are forced with trending popular contents instead of preferred and liked contents, attributing to "doomscrolling" where users simply scroll emotionally driving out content without any valuable information. This kind of experience leads to an exhaustion, eradication of trust on the platform and reduce user interest. Many individuals tend to other alternatives that match their preferences resulting in content bubbles that enforces biases or misinformation. By the end of this chapter, user will be able to

1. Appreciate the position of AI technologies in revolutionizing modern newsrooms
2. Identify the applications of AI in news generation, content building, personalization, and verification
3. Analyze the benefits and challenges of integrating AI into journalistic workflows
4. Examine ethical considerations surrounding AI-driven journalism
5. Explore how news organizations can maintain credibility, accuracy, and editorial standards in an AI-powered environment.

2 Related Work

Artificial Intelligence is transforming the news world into a new level. It is refining how the news is created, processed and delivered. It is also reducing the time taken for the journalists and the way it is consumed by the viewers. It has enhanced productivity and reachability. It has the above-mentioned benefits but also involves the following risk factors like ethics, authenticity and trust worthiness. This section of the chapter explores what it means to build an AI-ready newsroom, using the existing literature.

Impact of AI in all the domains is well known to us and it keeps growing exponentially. The literacy on AI is different from the other digital knowledge. The paper [1] is a review of sixty-eight papers in this domain. It reports that the AI literacy is still scattered and in the developing stage. The study includes the working of AI, its limits and its impact on the society. It also mentions about the people who use AI and for what purpose they use AI. It reports that more research is needed to adapt AI literacy for specific fields, like journalism, and different user groups. Generative AI's implementation into the field of journalism has caused a huge impact. According to the study in [2], generative models can assist journalists from news story conception to its delivery. Due to the implementation, there is a possibility of tension and those kinds of issues are discussed in this chapter.

The study reported in [3] discussed the latest breakthroughs in artificial intelligence and machine learning governing journalism, and whether AI will replace the employees. It also mentioned where AI is currently used, such as automated content generation, speech to text conversion, storytelling, translation, identification of false information, editorial assistance, and personalization. It also reported the technical foundation needed for AI implementation, algorithmic bias and equity in implementing AI. In today's world it is necessary to consider and study the details of the AI's impact across all fields. Journalism is a socio-technical domain, and AI has made a huge difference in this domain. The work reported in [4], addresses the problems of false information and trust issues in journalism using large language models. The work [5], reported an in-depth understanding of how AI has transformed journalism. In that work, one hundred and twenty-seven use-cases where chosen. There was a clear sign of increase in the use of AI. Mainly it was used for news writing automation, data processing and content personalization. If only AI is used without any human intervention, there is a loss of nuances and context in AI-generated news. They also reported the ethical challenges like algorithm transparency, data privacy and accountability. After analysing the positive and negative sides of AI in journalism it was also reported that there is a need for new kind of job positions like journalist programmer. In order to get the best in today's scenario AI competence is very much essential for journalists. The article gave a detailed analysis for regulating bodies, researchers and media personals. The article [6] discussed the advantages and disadvantages of AI in media. It exactly pointed out where the research gap exists and the solution for the problems. It correctly pointed out the areas where technical support is needed and where collaborative support is needed. With this approach, a responsible media technology can be devised for journalism.

The paper [7] is on the huge impact made by AI in English news translation. It also analyses what is the unique requirement in English News translation. It discusses in detail about data collection, processing, translation quality, and real-time translation technology. It also mentions about the future developments in AI based translation technology. The work reported by [8] analysed the news industry's adoption of AI across seven different fields: machine learning, computer vision, speech recognition, natural language processing, planning, scheduling and optimization, expert systems, and robotics. AI-driven news generation is mainly funded by major companies like Google, and only a few news organizations have access to this support. Among the seven subfields, machine learning, computer vision, and planning, scheduling and optimization are the most dominant.

The paper [9] discusses how AI will change journalism. Social media networks and mobile apps have already transformed traditional news. The sources of news have increased manifold. AI is expected to bring about a transformative change and usher in a new digital era. In [10] the authors discuss the application of artificial intelligence (AI) tools in the media and journalism sector, with a concentration on their application in content generation, editing, production, and dissemination. By using descriptive-analytical method, it summarizes that AI is widely used inside the domain of journalism. It also reveals the key challenges by experts, including bias, job loss, shortage of transparency, ethical concerns and deepfake content. In the article shown in [11], the study examines how misinformation is transmitted on Twitter because of AI-powered social bots. The harmful bot behaviour was studied by applying machine learning, actor-network theory and text mining on a huge number of tweets (around 30,000). The reported work gave a theory-based technique for understanding the functioning of bots in social media based false narratives. It provides a mechanism that assists organizations identify and address malicious bots before the occurrence of any harm.

The study reported in [12] investigates the new role of artificial intelligence (AI) in journalism in a developing country (Bangladesh). It demonstrates through surveys and interviews that, although Bangladeshi journalists show a positive attitude and a willingness to embrace AI technologies, their implementation is hampered by a lack of technical expertise as well as systemic issues like a lack of funding and infrastructure. This work emphasizes on the need for an AI-driven newsroom in a country and it also talks about the need for AI-based newsrooms in developing countries. The work also discusses about cognitive load needed by the media employees. The article reported by Nadiya Rahman in [13] also examines how the Bangladeshi Journalists are adapting themselves to the incorporation of Artificial Intelligence into the field of journalism. Though there are hurdles due to the native language and technology the journalists are incorporating AI into their work every day and they have shades of hope towards AI in the newsroom. This work was reported based on a study on fifteen professionals from media domain using Technology Acceptance Model.

Human-written content and AI generated news was compared and examined in the article reported in [14]. Based on the comparison, it was observed that there were numerous advantages like promptness and expandability. Along with the advantages, challenges were also analysed which included ethical challenges like algorithmic bias and concern to journalistic trustworthiness. It analysed the redesigned role of media employees in an AI-supported space, featuring the human-AI co-operation. Field study by significant news firms demonstrated practical situations. The authors concluded by addressing the upcoming trends, predicted patterns, workforce effects and the need to have responsible journalism in the age of AI. The study outlined in [15] stated that media personnel often tend to make intuitive decisions while choosing to use AI devices. In spite of the fact that they got advantages like better efficiency and data handling, there are issues on accuracy and reliability. There are also ethical concerns like bias. The work suggested for the need on algorithmic literacy, AI learning and continuous monitoring for responsible and ethical use of AI in newsrooms.

Based on the above articles, it can be understood that there is a common interface between AI and Journalism. There is a huge chance for growth in this interface. There are both positives and negatives, there is space for opportunities as well as concerns. The articles cited above highlighted the benefits and values that AI has added to the newsrooms. It also highlighted the drawbacks or threats, like ethical risk. With proper guidelines and precautions, AI will bestow for the development and reshaping of newsrooms and human supervision is central in evolving media newsrooms.

3 Methodology

A well-organized news production cycle (see Fig. 1) constructs valid news content from real–world incidents happening around us. The cycle begins with identifying and understanding the incidents happening around the corners, followed by idea creation phase where content developers build worthy news articles from contents. To improve the quality and reliability of the content, the concept was researched in detailed clearly understanding the relevant facts and details followed by validation of collected data. With trustworthy information available in hand the next stage involves creating the story, defining the structure and adding relevant multimedia content to it. During the next stage, the content is further reviewed and refined for more precision, correctness and alignment with the accepted editorial standards.

Once the story is finalized, it is moved into distribution stage where it is made available across major platforms like websites, social media or any broadcast outlets. The promotion phase confirms that the content reaches intended audiences through methods like search engine optimization, social sharing and push notifications. Once published, the content gets perceived by the public, sparking engagement through comments, likes, shares and feedback. This engagement triggers future ideation, there by completing the production cycle. In advanced newsrooms, all these stages can be supported and improved by AI to increase efficiency, audience impact and personalization.

3.1 AI Tools for Newsroom

The act of incorporating AI tools in modern newsrooms has transformed how the content is created, curated, perceived and delivered to the target audience. To achieve greater speed and accuracy in news delivery, AI can automate the routine for reporting and personalizing news feeds thus reshaping the journalism workflows. ML and NLP powered tools assists the journalist in real time fact checking, data summarization and trend analysis. With the support of AI systems support the audience are engaged by having content tailored to their specific needs, behavior and preferences. As AI evolves, it offers both the challenges and paths in achieving integrity, control and transparency in journalism.

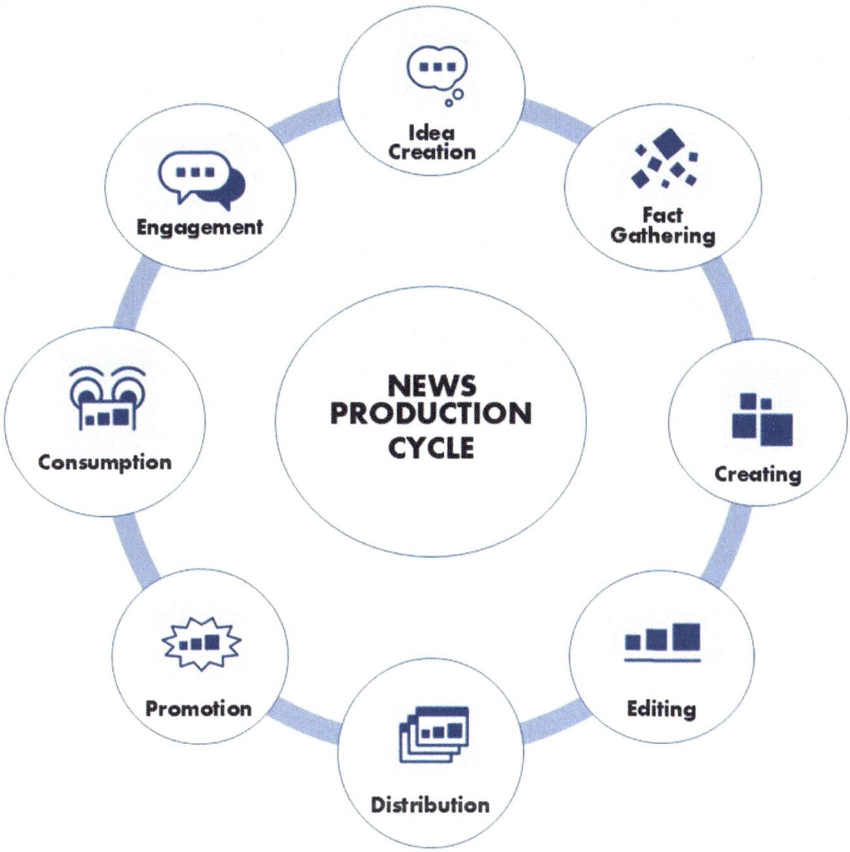

Fig. 1 News production cycle

The act of identifying premature signals of worthy news developments, contents, or sources before they get major attention is termed as lead generation in journalism. Lead generation tools supported by AI analyze huge amount of online content from social media, forums, blogs, press releases, etc. to identify growing trends, conversations, and breaking news.

Journalists using the lead generation tool will be able to snatch any real time happening before occurrence and stay ahead and can provide on time delivery of news. The lead generation tool catches up the trending topics, both in national and international level through the content engagement score which is usually very high for trending topics. These tools often fail to detect local news which is a valuable content despite poor online engagement score which proves to be valuable for specific communities. Local news rooms relying on lead generation tool may lose their strength in catching up local happenings and covering up local audience.

Present day lead generation tools assist the journalists in building admirable news content by grabbing valid timely clues from burgeoning trends. For example, Crowd-Tangle from Meta, happens to help the content developers by tracking how incidents get speckled across the social media. NewsWhip Spike analyzes in online how many users engage in content to underscore articles that are swiftly becoming popular. Dataminr leverages AI to transport real-time alerts about unnoticed events using publicly available data.

Google Trends showcases what people are searching for and reveals differences in interest across regions. Signal AI uses superior algorithms to track media coverage, identify topics, and analyze public feeling. Based on location, geo-specific data gives deep insights about trending discussion on twitter. To generate admirable leads, social management tools Hootsuite and TweetDeck follows the hastags and accounts and tries to derive the relationship among them to further strengthen the reliability of the news generated. Collectively these tools though they help the journalists to grasp clues earlier they fail to cover up the local worthy happenings which tends to be important for a local news room.

After lead generation, the next step is content creation. AI based content creation tools are revolutionizing the modern newsrooms by facilitating swift and scalable news fabrication. A streamlined news delivery process needs structured writing and news generation procedure. This is where AI tools come in action. From the output of lead generation, these tools generate final news by summarizing content from multiple sources. This news is then manually corrected to avoid errors and to cover contextual gaps.

Automated Insights' tool Wordsmith helps domain like sports and real estate in generating narrative contents to attract users. Narrative science's Quill generates earning reports, portfolio reports and analytical reports from structured numerical data by identifying trends and making comparisons.

Chatgpt support in content generation is by converting semi structured and structured data like statistics, weather, and stock information and turns them into readable narrative news automatically. Eg. HDFC Bank reported a 23% rise in half -yearly profit through strong mutual funds. Chatgpt can generate commentary, advocate headlines, provide summaries, render content, create ideas and redraft content. AX semantics, an NLP software platform built with facility for multilingual content generation and transformation supports journalism and reporting. Although these tools improve news delivery and productivity still human support is essential to validate and to ensure contextual correctness.

For news rooms to thrive in today's competition, real-time audience engagement tools have become an essential element to build stronger customer relationship, retention and personalization of news delivery. Journalists have to understand how the content delivered is working and how to further tune them as per reader's choice. Parse.ly and Chartbeat are digital analytical tools for journalist that help publishers and newsrooms to optimize the news delivered through analytics.

For automated newsletters based on the segmented audience's preference, tools like Hubspot and Mailchimp segment audiences based on metrics like conduct, significance, time spent and click rate. To encourage two-way communication between newsroom and reader Disqus collects reader comments, creates dialogue between them. Piano manages subscriptions and personalized delivery based on reader preferences and past history.

Any news needs to be delivered to mass audiences for successful business and hence need a powerful distribution tool. Google translate, DeepL, Microsoft translator helps journalism by automated translation across multiple platforms and formats ensuring contents reach global audiences in few seconds. Descript and Lumen5 can convert text into storytelling formats like videos and podcasts.

3.2 *Functions of AI Based Newsroom*

Please note that the first paragraph of a section or subsection is not indented. The first paragraphs that follows a table, figure, equation etc. does not have an indent, either.

Listed below are the six main functions of a futuristic AI powered newsroom, depicting how AI improvises all aspects of future journalism (see Fig. 2).

- **Automated Content Creation**: AI based tools using Natural Language Generation creates news reports, catching headlines and customized summaries. These systems enhance the procedures involved in content creation with a rational style and tone helping the content developers to center their works towards deeper investigation and content quality improvement.

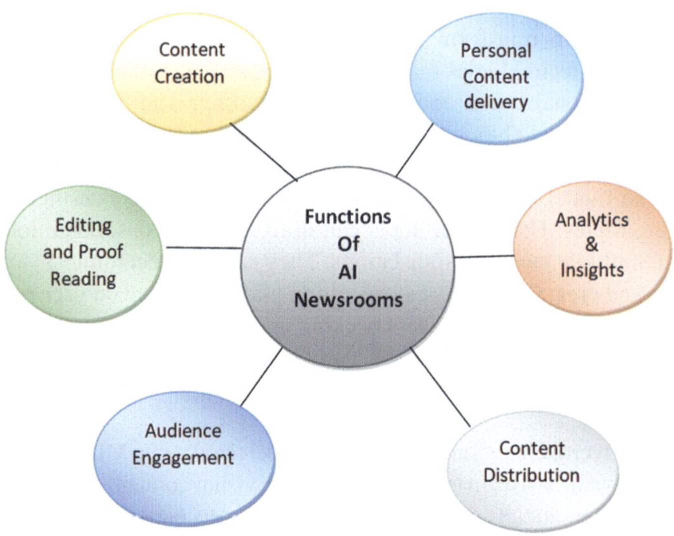

Fig. 2 AI newsroom functions

- **Editing and Proofreading**: AI based grammar checking tools assists the content developers to increase accuracy, format presentations and enhance legibility of the content. These tools help in proof reading, identifying and correcting errors, redefining languages, maintaining editorial ethics and standards, guaranteeing flawless publication content with reduced manual intervention.
- **Personalization**: By investigating the customers reading behaviors like click patterns, reading time, search history, preferred content and visualization methods AI can customize the content to be delivered. This in turn will enhance user engagement with media, user contentment and promotes retention across all type of platforms.
- **Analytics & Insights**: AI inspects real time parameters like number of visitors, bounce rates and engagement patterns and levels. It aids the newsroom controllers to understand how audience behave, their trends and guides controllers in making informed decision based on data analytics about the content and user.
- **Content Distribution**: As user's views news articles across different platforms like smart TV, mobile devices, web pages and social media, AI can optimize the formats, style and timings according to the user type and device specification.
- **Audience Engagement**: Feedbacks collected from users through any AI based tools like chat bots, virtual assistants, survey platforms, sentiment analysis tools, conversional feedbacks and real time feedback loops helps refine the content as per user expectation and foster high level of engagement.

3.3 Key Technologies Driving AI Based Newsroom

AI basically aims to transform the journalism process through a collection of advanced enabling technologies that promotes streamlining of tasks such as content curation, data personalization and preicision reporting. The key AI technologies, from machine learning to block chain enables how future newsroom can be powered to meet standards and stay in competition (see Fig. 3). Using machine learning the system can model breaking news, detect fake news using classifiers and identify emerging topics. Using NLP tools can draft articles, consolidate long reports and perform multiple language translations. They can also recommend stories, push notifications and create interactive story telling plaforms based on user preferences. ML tools can automate task like fact checking and verification, tagging, categorization and archiving of articles based on many labels. By using ML models, the ads can even be customized, read customer minds who like to withdraw and maximize the subscriptions.

Fake news, advertisements and images can be detected using ML and DL methods. Computer vision can assists in tagging elements in visual content, detect fake content and may also automatically create video summaries to enhance better delivery of news. Using speech processing tools AI can convert interviews into text articles and permits voice assistant to deliver real time updates of events enabling interaction sessions and explore in depth about the events. Deep learning can further enhance

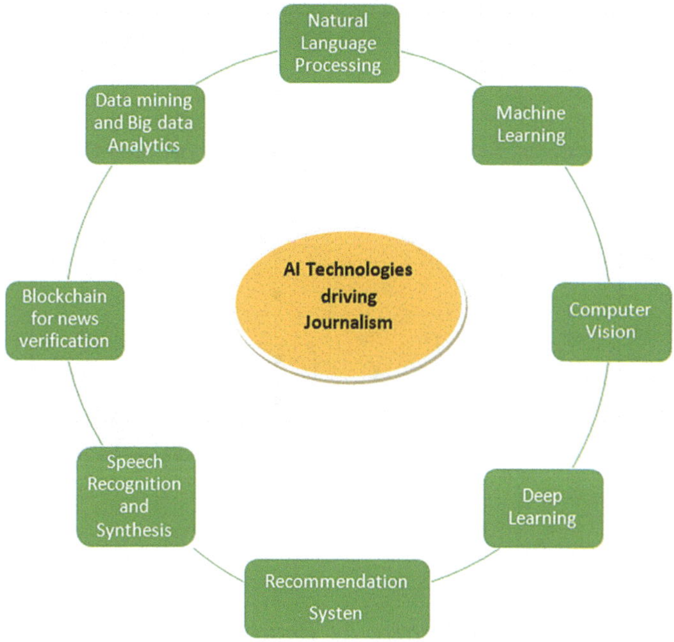

Fig. 3 Newsroom 3.0 enabling technologies

the new delivery through real time video editing. For example, the Associated Press uses AI tools to generate financial reports from structured data automatically and aids the reporters to focus their attention to further investigations. Most happening updates about games, wars, elections and weather can be automated without any human intervention.

Block chain can be used in automated journalism to ensure the authenticity of content sources, origin of contents reducing misinformation and ensure the security of the content ownership. Most misinformations/fake news use exaggerated language, sensational words or title. Using ML/DL model AI can analyze the style, sentiment, tone and emotion used for creating fake articles. Cross checking of facts like dates, numbers, values or statement can also be automated using NLP. By flagging of misinformation sources, evaluating news outlets, authors, domains, credibility of data sources, identifying propagation patterns, account boosting, misleading stories AI can build a powerful news room in future. All together new technologies reshape how journalism functions making it more efficient. The graph illustrates the dramatic shift in usage of AI based tools between 2015 and 2025 (see Fig. 4).

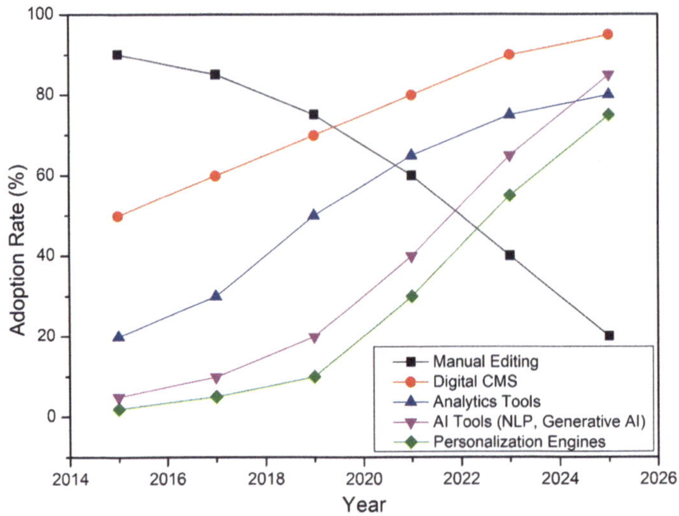

Fig. 4 Trends in newsrooms

3.4 Working Model for Future AI Ready Newsroom

To understand how AI fits into modern newsrooms, it's useful to view the news workflow in three key stages:

- News generation—involving research, source discovery, and topic selection.
- Production—where the story is written, edited, and refined.
- Distribution and promotion—which includes sharing stories, optimizing for search engines, and crafting content for social media.

The architecture of AI supported future newsroom consists of various elements (see Fig. 5). The content for the news article is collected from multiple sources like web pages, social media ranking and scoring system feeds, web crawlers and APIs. The content is then processed using methods like data cleansing and structuring. Processed data is then fed into an NLP engine that generates structured data which are then archived in news stories in specific formats. For example, relational databases like Mysel, Postgre are used for storing data about subscriptions, articles, user accounts and tags along with metadata. NOSQL databases for faster retrieval of multimedia data.

Parallelly a user profile monitor tracks user patterns in clicking, timing and interactions and builds a user profile. These insights given as input to user profile generator then customizes the content delivery. The personalization engine then applies various techniques like content-based filtering, reinforcement learning, rule-based filtering, demographic and behavioral personalization on generated profile to get a complete recommendation for a given user profile and customize news delivery. Recommendations are then ordered, ranked and scored to improve their quality and relevance.

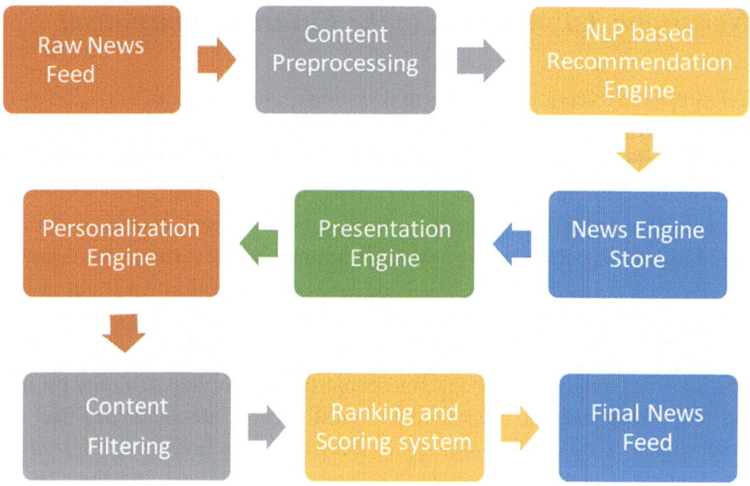

Fig. 5 Architecture of AI-powered news delivery platform

Finally, the news is presented to the reader through an appealing interface ensuring a reading experience solely based on user preferences, style and past engagement. This loop promises accuracy and 100% customized news delivery over time, periodically refining the way news is delivered for each user,

The depicted architecture can be applied to deliver live score updates for a cricket match to engage users actively off ground and give a personalized experience of enjoying the match virtually. Firstly, updates about the match collected from various sources as mentioned earlier are processed then formatted and filtered as per the need extracting crucial information like scores, stats, locations, achievements and histories. Processed and filtered content is then passed through a content representation engine where the data is structured and tagged using NLP methods like tokenization, stemming, named entity recognition, part-of-speech tagging, keyword extraction and sentiment/Emotion analysis.

At the same time, a user profile is generated and stored based on how the user engages with the match, what content he is interested in and how he interacts. The personalization engine then identifies and maps a personalization technique highlighted earlier and updates the user profile and prepares content accordingly. Using customer feedback and applying reinforcement technique the architecture ensures to continuously improve the content and delivery. The ranking and scoring module based on personalization engine output sorts the content and arranges as per the user preferences highlighting trending and relevant information for individual users. At last based on user preferences in terms of language, location and device specifications live updates are personalized and delivered through personally customized dashboards that effectively engages the audience.

3.5 Newsroom 3.0: Adapting to the Future of Journalism

As an AI-supported version Newsroom 3.0 binds various entities like on-spot data coverage, digitalization and data analytics. Basically, AI extends its support in terms of customized news generation, assisting in content development, article writing and customizing content publication for different digital platforms. AI supported newsroom tends to build a smart newsroom by improving article relevance increased engaging of users and enhances the overall efficiency in delivering news content. Any AI powered newsroom adopts the following approach in building a smart newsroom.

Understanding User:

The first step in personalized delivery is to construct a complete dynamic profile of all customers capturing both the interested and indifferent contents as each user is likely to have varying interest in different topics over time. Techniques preferred include seeking precise choices, capturing the hidden indirect actions and their feedbacks. For example, directly asking their topic of interest, recording the sites visited, ads clicked, time spent followed by the response for each content like ads not interested, seen many times etc.

For example, as an occasional news reader if the user repeatedly browses through recent trends like weather forecasts, increasing gold rate and film fare awards then the AI powered newsroom is empowered to give to updates like

- Amount of rainfall in Kerala's Idukki district, gold rate at different regions.
- Generate a comparison of rainfall over a decade or trends in gold prices.
- Recommend best season to visit Idukki or buy gold.

The same content can be customized in different languages and delivered to different users based on their preferences. This level of personalized content makes the news to be globally responsive, enhancing user commitment to newsrooms across boundaries.

Processing the Content-News Understanding with NLP:

For an effective news delivery two key elements are users and content. The extent to which the user is understood and the content can be made engaging is very important. Natural Language Processing algorithms extracts essential features from every article like:

- Focus (e.g., war, technology, medicine)
- Entities (e.g., "gold", "Richard", "Bank")
- Reaction (e.g., sanguine, catastrophic, impartial)
- Readability level, language, and even tone.

Transformer based models converts the article contents into dense vectors of words called as embedding that captures the right spirit of the content like the relationship among words which in future helps AI in delivering a quality content.

NLP enlarges the cricket coverage by providing live data and transforming into personalized summaries and matters. For example, when a player scores century, NLP automatically constricts highly contextual updates like, "Player hits excellent century against Australia off in 75 balls".

NLP extracts diverse information from various sources like match summary report, news articles, social networking platforms, blogs, comments, meetings and conferences. Factors like sentiments, emotions and critical moments are extracted as per audience preferences and multilingual reports can be generated.

Besides all the above, NLP can analyze the audience preference based on favorite cricketer, famous matches and memorable events and arrange the report and structure it in a manner that engages the audience effectively.

Generative–AI creates a new era in journalism through snappy stories, fresh story telling techniques, visuals and methods. Reel Framer, a human-AI collaborative and creative system assist journalist to transform text into short simple videos and story boards to render an informative and entertaining content to the audience.

Finally, NLP associates raw event data and user entertaining news content, enabling swift, smart, and customized sports journalism in the period of AI-based newsrooms.

The Recommendation Engine:

The suggestion engine is accountable for suggesting and crafting audience specific news content. The core engine then matches the user profile with the news content. Two prime techniques used here are:

- Collaborative Filtering: Suggesting content based on what similar users enjoyed
- Content-Based Filtering: Suggesting articles similar to what the user already likes.

Advanced systems combine both filtering approach, balancing newness, miscellany and personalization. Additionally, with reinforcement learning the model can adjust the content over time based on user engagement, feedback and satisfaction.

Ranking and Score System:

The Ranking system calculates scores from personalization processes and assigns the scores to all indexed news for each user. The highly scored news is chosen and delivered to the user to ensure personalization based on metrics like relevance, time and content quality the ranking and scoring systems can arrange and personalize news content. Using NLP based ranking the content can be ranked according to user characteristics.

- Content Scoring:

  For any kind of cricket related news each content is given a score based on relevance based on keywords, engagement, sentiment and freshness. For example, a tag line like "Indian Women's team won world cup tournament for first time" might get a higher score than a general story like "Indian Women Team wins" because it is very focused and emotional.
- User Profile Matching:

The AI ranks articles based on how well they align with user interests like much loved teams, Key players, ideal format and past behavior.

- Dynamic Ranking:

  As a match progresses, live events improve the rankings. Under ranked articles can be removed and new articles can get higher score.

- Multimodal Output:

  Depending on the score the news can be updated to cover different users and different forms of delivery like reels, comments, infographics or notifications can be made.

Feedback Loop—Adapting to Trends:

Any personalization system that adapts through continuous learning always stands unique. From the users feedback the process repeats where the articles are re-ranked, contents are either adjusted or modified and the emotional connections are improved either increased or decreased.

For example, if a user suddenly avoids political news and prefers science features, the system understands this shift and cleverly adapts the news feed. Suppose a user initially engages with articles about World Cup highlights and specific players' performances, so the AI-driven newsroom shows more content related to matches and legendary players.

However, over time, if the user begins ignoring world cup updates and prefers to have more on Test match analysis then the feedback Loop processes again and re-ranks the article covering recent information and modifying the content based on various choice and adapts a dynamic sensitivity to content.

Evaluation:

Finally, to grow continuously and to be key player for long time evaluation is an effective tool to assess the growth and effectiveness. Hence new kinds of benchmarks are to be identified and put into action for evaluation. First of all, the system has to understand how the news reaches the user, for example through websites, search engines, social media apps, newspapers, pamphlets or other outlets. This insight helps assess the visibility and discoverability of journalism. After the user access the content AI monitors the user interaction in terms of intensity, scrolling behavior, click patterns and content sharing. Using this we can evaluate the content reachability and enhance the audience engagement quality. Content preferences serve as a valuable insight for editorial section to plan for future content delivery.

Location based services assists in improving the reachability of content across varying communities in terms of time, speech and culture. Standard thresholds need to define to measure the time difference between AI based content generation and delivery. Hunting down the amount of content published over the period of time either daily, weekly, yearly or monthly offers a gauge of newsroom productivity. Finally comparing the similar size and type newsrooms in your community or location will give valuable contextual clarity of the process. Self or peer benchmarking guarantees to uncover the potential points for growth in terms of outreach, novelty and services.

Ethical concerns:

AI tools provide us a valuable support in terms of user engagement and business improvement and at the same time raises concerns in terms of ethical standards followed. Most of the AI tools rely on personal data analytics which can compromise the privacy, security and transparency of the user. Mostly the customers are unaware of what data is collected, who is collecting, how are they analyzed and stored. Sometimes the trust on the media is weakened when limited access to user data results in echo chambers where only restricted /limited user data is exposed and used for content delivery. Even though through automation news delivery has fastened and scaled up but it is not flawless. Auto translations may misinterpret quotes, references, cultural information and tones and end up in generating inaccurate and sensitive content.

To ensure cultural context and linguistic correctness, human intervention from experienced and professional translators is necessary. When a short video clip is generated from a long investigative report using AI there are chances that the video may misinterpret or oversimplify facts raising the ethical concerns. The ethics becomes a concern when correctness and context are lost during repurposing. AI personalization risks where viewers are exposed to restricted views that reinforces certain beliefs reducing the trust on the newsroom. Sensationalized content from repurposing can sometimes undermine the trust on the media. AI hides the truth about whether a story is custom build by humans or generated by algorithms and hence losing its transparency.

4 Challenges in Newsroom 3.0

- Bias

  AI models based on biased data algorithms can reflect skewed coverage of news. For example, if the training set used concentrated on male cricketers or international players, then unrecognized regions, female cricketers or local players may get less attention, promoting gender inequality in sports which is results in a dangerous environment.
- Accuracy

  When newsroom 3.0 depends solely on AI for various stage of news development like editing, summarizing and reporting it may sometime misinterpret some complex statistical information or inference in the game giving misleading stories which eventually needs an expert to handle the situation.
- Overreliance

  Overly reliance on AI based journalism can reduce the real journalism standards. For example, dependency on AI generate post-match analysis of games may undermine skilled players, their on-field emotions or techniques used that only a human expert can identify, analyze and appreciate.
- Sensitive Reporting

AI-generated content can mishandle situations like retirements, injuries and controversies among the players as it fails to apply human ethical context to handle the data. Such kind of reports may look insensitive, rough and risk the reputation of a player.

- Human Involvement

  Any kind of AI-generated news content needs to be manually corrected, the context improved and ethical values upheld and trust and fairness ensured and cultural warmth promoted in every story that is being shared.

- Ethical Concern

  On a wider angle, the factual accuracy of AI generated news is to be tested to avoid any misinformation being circulated. However, it is mandatory to ensure the quality of service and ethical standards remains up to the standard. Pillars of journalism that is correctness, integrity and openness of the sensitive data have to be maintained irrespective of the AI methods used. Well informed editorial insight has to be ensured to preserve the quality of AI generated content.

5 Benefits of AI Newsroom

- Increased Reader Engagement: AI based personalized delivery in forthcoming years will enable users to experience news being delivered to their platforms based on his/her preferences, reading pattern and interactions. By providing relevant and favorite news articles along with multimedia effects can effectively improve the user experience and engage the user for a longer time based on his/her needs giving him/her a favorable and convenient environment.
- Enhanced News Consumption Experience: AI promotes a more individualized and effortless experience for news readers. By evaluating the customers behavior and content expectations, AI newsrooms can generate new news topics that are in track with the customers preference, enhancing customer satisfaction and motivating the customer to make frequent visit to the news portals.
- Improved Subscription and Conversion Rates: Customized delivery of news content likely to convert a visitor of news portal to become a regular customer or subscriber increasing the income for news agencies. By recommending articles or news as per the requirement of the customer the journey can convert a normal user to get paid subscriptions further enhancing the attachment with the portal.
- Stronger Reader Loyalty and Retention: Regular and appropriate delivery and invaluable content create loyalty among the news readers. It builds a kind of trustworthy bonding and promotes user engagement for long time and achieves higher user retention rate which clearly records the success of the portal.
- Actionable Audience Insights: Many AI tools collect the trends, patterns, preferences and interaction patters of the users and based on that future newsrooms can envision the customer needs and refine its plans, methods and ensure that

they retain the customer for long period and give him a satisfied and compelling experience.

- Tactical improvement in a Competitive Market: As the AI based news system grows linearly staying in market and retaining the content delivery quality is a challenge that needs to be addressed. To cover the entire population and improve business integrating AI into news delivery is a mandatory task.

6 Performance Evaluation

As AI based journalism is the need of the hour there has to be a structured procedure to evaluate the performance of the newsroom activities. Complex algorithm backs up the AI newsroom process from lead generation to publishing, the future of these AI powered news rooms relies on various technologies, its strength and power and hence they have to provide an expected performance to stay in market.

To identify any kind of slack in areas like content delivery, resource usage, user experience or incorrect information delivery it is necessary to evaluate the performance of the system to prove its worth and effectiveness in news delivery. Performance evaluation helps to identify bottlenecks, boost the system capability, guarantee ethical compliance and build trust among the audiences.

- Latency Reduction

  Latency is defined as the time elapsed between the occurrence of worth news to the time the system generates content and updates. Latency can be experienced in all the phases of news development from data collection, NLP, proof reading to content delivery and distribution. For real time events and critical news coverage minimizing latency is essential to stay competitive and relevant in delivery news. To stay alert, receptive and ahead of competitors it is essential to ensure that the system is a low latency system.
- Response Optimization

  Response optimization evaluates how effectively the AI system can handle requests from users like editors, journalists, regulatory agencies and end users for searching, reviewing and validating auto generated data. The system should be capable enough to deliver correct results quickly and manage any request during peak hours and to prioritize the delivery based on the demands. To maintain a fine editorial workflow, it is mandatory to stream line the response time for delivery of news to the users.
- Bandwidth Optimization

  In future most of the device's carried by the audiences will be small, portable, low powered devices for easy access and mobility. Bandwidth optimization refers to the ability of the news delivery system to provide high quality multimedia content like video, audio, graphics and interactive elements using minimum network elements. Techniques like compression, encryption and various content management strategies can guarantee seamless delivery of content and experiences across diversified devices and locations.

- Resource Efficiency

  Resource efficiency is measured in terms of how AI tools consume power and memory. AI enabled tools are mostly in need of large amount of processing capability (CPU & GPU) and storage space (RAM). Any resource efficient tool providing large scale news delivery needs to provide quality service with lower environmental foot prints, reduced functioning cost and minimum wastage. Resource efficiency also means the support extended to scale the operations without affecting the existing infrastructure.

- User Experience Enhancement

  Finally, the success of an AI assisted newsroom is observed from how users can engage with the content delivered. Metrics based on user engagement includes the content personalization, effortless navigation among content, quality of multimedia data and the valuable time spend. A positive user experience drives audience retention, trust, and content sharing. AI systems should be evaluated for how well they tailor content to individual preferences, adapt interfaces for accessibility, and maintain transparency in content curation.

7 Conclusion

Similar to social media platforms, AI has also transformed the way by which customers consume news by personalizing the content in terms of their interest. Popular platforms like Netflix, Facebook, Instagram uses large amount of data points to analyse the patterns and behaviours and aims to provide contents customized for individual user. AI traces all the paths of a user starting from what type of content they prefer, time spent on each article, their feedback, articles they avoided, comments made etc. By integrating AI, Newsrooms have transformed rapidly into a smart and dynamic industry. AI thus empowers the newsrooms to deliver creative customized contents, conduct deeper investigations and improve the efficiency of the newsroom process. Though AI as a technology serves human kind in engaging them with needed contents it is essential to maintain ethics, transparency and emotions. To conclude AI newsrooms not only serves as an alternative for journalist but also aims to increase the potential of news industry in the era of information.

Declaration of Generative AI and AI-Assisted Technologies in the Chapter Preparation Process During the preparation of this work the author(s) used ChatGPT in order to collect and organize relevant information. After using this tool/service, the author(s) reviewed and edited the content as needed and take(s) full responsibility for the content of the published article.

References

1. Pinski, M., Benlian, A.: AI literacy for users—a comprehensive review and future research directions of learning methods, components, and effects. Comput. Human Behav. Artif. Humans **2**(1), 100062 (2024)
2. Nishal, S., Diakopoulos, N.: Envisioning the Applications and Implications of Generative AI for News Media. arXiv preprint (2024)
3. Kevin-Alerechi, E., Abutu, I., Oladunni, O.: AI and the newsroom: transforming journalism with intelligent systems. J. Artif. Intell. Mach. Learn. Data Sci. **3**(1), 1930–1937 (2025)
4. Cheng, S.: When journalism meets AI: risk or opportunity? Digital Government Res Pract **6**, Article 10 (2024)
5. Sonni, A.F., Hafied, H., Irwanto, I., Latuheru, R.: Digital newsroom transformation: a systematic review of the impact of artificial intelligence on journalistic practices, news narratives, and ethical challenges. J. Media **5**(4), 1554–1570 (2024)
6. Trattner, C., Jannach, D., Motta, E., Costera Meijer, I., Diakopoulos, N., Elahi, M., Opdahl, A.L., Tessem, B., Borch, N., Fjeld, M., Øvrelid, L., De Smedt, K., Moe, H.: Responsible media technology and AI: challenges and research directions. AI Ethics **2**, 585–594 (2023)
7. Lin, J.: Artificial intelligence applications in English news translation: strategies and research. Modern Manage. Sci. Eng. **6**, 140 (2024)
8. de-Lima-Santos, M.F., Ceron, W.: Artificial intelligence in news media: current perceptions and future outlook. J. Media **3**(1), 13–26 (2022)
9. Broussard, M., Diakopoulos, N., Guzman, A.L., Abebe, R., Dupagne, M., Chuan, C.-H.: Artificial intelligence and journalism. J. Mass Commun. Q. **96**(3), 585–615 (2019)
10. Aissani, R., Abdallah, R.A.Q., Taha, S., Al Adwan, M.N.: Artificial intelligence tools in media and journalism: roles and concerns. In: Proceedings of the 2023 International Conference on Multimedia Computing, Networking and Applications (MCNA), pp. 19–26. IEEE, Valencia (2023)
11. Hajli, N., Saeed, U., Tajvidi, M., Shirazi, F.: Social bots and the spread of disinformation in social media: the challenges of artificial intelligence. Br. J. Manag. **33**, 1238–1253 (2021)
12. Partha, S.B., Tabassum, M., Goni, M.A., Kundu, P.: Artificial intelligence (AI) and future newsrooms: a study on journalists of Bangladesh. Pacific J. Rev. Te Koakoa **30**(1–2), 96–110 (2024)
13. Rahman, N.: AI in Newsrooms: Assessing Perception, Challenges and Possibilities in Bangladesh's Media Landscape. Ph.D. Thesis, University of Memphis (2025)
14. Friday, J.P., Soroaye, M.P.: The Rise of AI Journalism: How Algorithms are Shaping News Content (2024)
15. Cools, H., Diakopoulos, N.: Uses of generative AI in the newsroom: mapping journalists' perceptions of perils and possibilities. In: Digital Journalism (2024)

The Future of Journalism in an AI World

Lekha Kannappan, Geetha Anbazhagan, and Aruchamy Prasanth

Abstract Rapid development in Artificial Intelligence tremendously influences the field of journalism, which causes potential interesting challenges for the sustainability of news media. The chapter showed the impact of AI in journalism in connection to business systems and their inferences on ethics, public discourses and democracy. Secondly, basic functions such as collecting verified news, its analysis and distributions of the journalism are tested for the effect of using AI. In addition to details on AI media workflows, capabilities now required by newsrooms and details on the changing nature of newsroom operations in AI-enabled media, a comprehensive section is also dedicated to the ever-evolving role of human journalists in AI-enabled media. The report concludes with practical recommendations for journalists, the media sector and policymakers on how AI systems can be used for democracy in journalism. In the end, there are suggestions for journalists, media houses, and policymakers so that they can make use of positive sides of AI for journalism, which help to enhance democratic institutions.

Keywords Digital era · Artificial intelligence · Big data · News rooms

L. Kannappan
Department of Electronics and Communication Engineering, SRM Valliammai Engineering College, Kattankulathur, Tamil Nadu, India

G. Anbazhagan (✉)
Department of Electrical and Electronics Engineering, College of Engineering and Technology, SRM Institute of Science and Technology, Kattankulathur, Tamil Nadu, India
e-mail: geethaa2@srmist.edu.in

A. Prasanth
Department of Computer Science and Engineering, Vel Tech Rangarajan Dr. Sagunthala R&D Institute of Science and Technology, Chennai, Tamil Nadu, India

P. Chien Thang and A. Prasanth (eds.), *AI-Assisted Journalism and Media: Opportunities and Challenges*, Studies in Computational Intelligence 1277,
https://doi.org/10.1007/978-3-032-19648-4_10

1 Introduction

Use of AI has led to a huge advancement in journalism recently. With the application of AI, even the part of the news industry such as how materials are created which are visible to users who read them and analyze them has changed [1]. With automated technology, news organizations can write more stories about financial reports, sporting events, and weather. With this technology, journalists can pursue more difficult news stories while AI is able to cover the simple ones. With machine learning, news delivery systems can analyze people-favorite content and use it while delivering content that is more likely to attract readers.

Now, thanks to AI, it is much easier for journalists to investigate and analyse big data. In contrast to what a single data source can find the data visualization can reveal some important patterns and insights difficult to point out and report on before by generating detailed enlightening content on challenging issues [2]. However, the introduction of AI in journalism raises many ethical issues that should be aware. As AI continues to develop at a faster pace, more people are concerned that computers will be able to take over some of the roles that journalists carry out, particularly in places where a lot of content is created automatically.

Many experts find that AI systems are a major cause for concern due to the hazard of algorithmic bias. Because these systems use data that already exists, any existing biases in the data can be passed on to the system. Having these biases leads to the possibility that AI reporting will also be inaccurate or unfair. In addition, there is a serious issue related to possible automated circulation of fake news. As machines become better at writing in human ways, the chance of them being used to create or share large amounts of false information increases. Because of this threat, the faith that many have in the news and journalism are seriously at risk.

Journalism groups need to develop useful and practical guidelines for AI, that everyone agrees are ethical. They need to maintain the highest moral standards of journalism and make it clear when content is generated from AI. Training journalists on the ethics of AI in their work should occur all of the time. Besides, for AI systems to adhere to the journalistic values, it is necessary to bring together technologists, ethicists and journalists [3]. By bringing together expertise from multiple fields, it can make sure AI in journalism is a supporting force and not a challenging force to the positive reputation of journalism. Energetic efforts need to be made to ensure that journalism does not lose its crucial characteristics of accuracy, fairness and ethics due to new technological advancement. This way, what current journalism should do is to benefit from the powers of AI and keep alive its role to disseminate and distribute significant pieces of information for society.

2 The Evolution of Journalism in the Digital Age

There is a lot that has changed in the way that news is received since the use of internet and mobile devices made their debut in society [4]. Nowadays, news is shared and obtained through websites and social networking, leading to a change in the way people access, engage themselves with, and relay news. The development of internet use in late 20th, the early 21st year hugely parted the way news information for being offered from the previous styles. Unlike traditional newspapers, radio and television, news websites are available all the time and have unlimited space to publish stories. News groups established themselves on the web which made it possible to report news instantly and updated articles right away [5]. Now, in the online media, people can obtain news from various parts of the globe without delay. Figure 1 shows how media and communications technologies have evolved beginning with print newspapers in 1605 to today's AI and Augmented Reality (AR) chatbots in 2010.

Next smartphones and social networks, citizen journalism means more people are receiving and delivering news which many times beats old media in bringing news-worthy happenings to our attention, especially when it comes to timely news contents. Also, changes in how it consumes and share news have altered how the industry works financially as well. Many different online-only or online-first services have started-up, but these innovations are difficult for traditional newspapers and magazines. With the move to digital media, publishing companies have introduced paywalls, targeted their advertising and relied on sponsored articles to generate money from their websites [6].

Because of digital transformation, people's understanding of and use of news has been deeply affected [7]. Because of the plethora of sources and viewpoints available to us readers have to work harder so they will be able to make sense of what they encounter. Personalized news feeds are raising warnings about people only seeing a limited selection of opinions. As a result, using internet and mobile technologies,

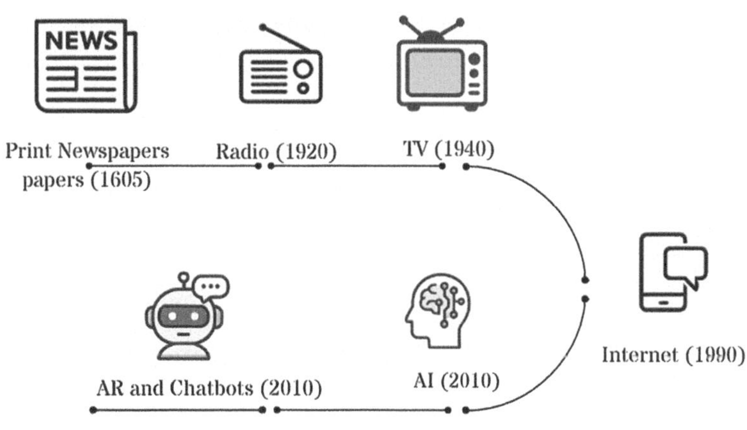

Fig. 1 Evolution of media and communication technologies

people now receive news instantly, can access it from anywhere and can become more involved. Even though everyone can now access information more easily and a larger number of voices are heard, such changes have also led to questions about the quality of news, the ability of the public to judge news articles and the future of journalism [8]. With the evolving digital world, news consumption and production will very likely evolve in directions that are not yet visible to us.

2.1 Modern Newsroom Challenges

Because of digital disruption, news organizations have had to change the way they operate, conduct business and deliver news [9]. Several things contribute to this disruption such as ongoing changes in what readers prefer, fresh technological innovations and developments in how publications earn money. Nowadays, readers expect news quickly, want content tailored to them and interact with it, using their phones and social networks as their main ways to get news. Because of this change, news organizations now produce easy-to-read content, update their websites for mobile users, use social media, apply algorithms to recommend stories and share news in novel ways [10].

A reduction in advertising revenues has strongly reduced the traditional earnings of news organizations. Switching to digital publishing has affected advertising and even though more ads are being placed online, print ads are still being cut back sharply. As a result, different news organizations have experimented with paywalls, asking readers to subscribe, adding advertising from other sources, posting ads that look like normal news articles and engaging in other profitable activities.

Competition from new media sources has led traditional news companies to upgrade their content with novel report styles, different storytelling approaches and joint efforts [11]. They have put a lot of effort into building a well-recognized brand, being true to facts and keeping up with misinformation. In recent technology, news organizations must now rely on AI, Virtual Reality, AR, voice-activated gadgets, blockchain and big data uses for producing, distributing and analyzing their content and audience [12].

Based of digital changes, newsrooms are being adjusted, and journalists need to learn new skills. Organizations in media have started reporting online first, giving journalists skill in multimedia and hired specialists with different strengths to promote experimentation and advancement [13]. In the information age, ethical issues have become more important, so news organizations have developed rules for social media use, controlling speed with reliability, dealing with user-generated content, information privacy, possible algorithmic influence and guarding editorial freedom as they look for alternative funding [14]. In essence, digital disruption has urged news companies to look at fresh ways to bring out their content, share it and earn income from it. Adapting successfully means using new technologies, finding several ways to earn income and staying ethical in the growing media industry.

3 Realizing AI in Journalism

Machine learning, natural language processing and computer vision are the main tool of AI uses in journalism. A variety of applications make use of these technologies such as analyzing data, creating content automatically, understanding people's views, automatically translating, confirming visuals and providing custom news. With AI, investigative work becomes more effective, monitoring can happen in real time and checking facts is enhanced [15]. Yet, bringing AI to journalism raises doubts about bias in algorithms and privacy. The role of AI in transforming journalism does not change the fact that it needs people who follow basic journalism rules. AI is used to increase human expertise rather than to replace people completely [16]. The increasing use of technology requires news media to keep journalistic standards high in its future.

3.1 The Role of Big Data in AI-Powered Journalism

Today's news and content depend on AI in significant ways because of large data collection and analysis. Here, data is gathered from various sources and then utilizing AI, the team notices similarities among the data points. Public opinion and information distribution are investigated by using algorithms on social media posts, comments and articles [17]. Computers and software are used to forecast what may happen next and language-checking systems measure users' feelings. These systems are including information validation systems, content recommender systems, and data mining reporting. In this field, many computers sifting through many amounts of data are used to detect potential misconduct. However, there are still concerns surrounding the use of AI, including privacy issues and the potential for biased outcomes and the spread of inaccurate information. For AI to be effective, the media must stay fair, accountable and ensure an adequate role for humans to keep the application of AI from eroding journalistic values. As AI technology advances, it is likely that trend and insight discovery will continue to be facilitated by AI for news. One issue for media organizations is how to balance automation analysis with rational editing, ethical care and respect and care for individuals in reporting.

3.2 Types of AI Systems in Newsrooms

From automating content curation to personalizing the content according to customer preferences and from optimizing resources in the news organization to enhancing the delivery of news, AI tools have become an integral part of the news organization's operations. There are three primary types of AI solutions, which can be roughly categorized in the following manner:

Automated Content Generators

Machine learning is being used to create written content automatically. They can apply the data and sources retrieved from the news to generate articles, summary reports. For example, these generators help tremendously because they share a lot of numbers (such as data from stocks, scores for sports events and reports). They enable stories to be available in many languages and provide the latest coverage of news stories. However, anyone's systems must be inspected to find out how well, how accurately, how timely and how well the information they generate is relevant to each person's needs.

Recommendation Engines

Journalists used smart algorithms on user data, preferences and past reading habits to recommend what is most suitable for each user [18]. They find out how users interact, so they know which topics are more likely to catch their attention. With personalized news, they increase the chance users will return, help choose where news is displayed online and update targeted ad plans. As a result, recommendation systems help news outlets enhance reader activity, extend the time users stay on the website and improve their financial performance by selling better content.

Data Visualization Software

By using AI, news organizations are able to present detailed information using techniques that all readers can understand. They can instantly create charts, graphs and infographics from stored data, allowing readers to interact with the visualizations and identify trends that are harder for people to see, create visualizations in real time to report on breaking events, plus improve storytelling by making harder concepts clearer. Journalists usually utilize various data visualizing tools to report different news so that it makes the public ease to understand [19].

The journalism field holds a great impact among the newsrooms by using AI technology. It helped the journalists to focus more on the interesting sectors of the news. They let news organizations create additional content and inspire more readers. AI aids to guarantee that the content delivered in news is tailored for every individual, serving them to uphold their interests and engagements within the news offered. Thus, AI in content generation makes it possible for editors and business leaders to make choices of their strategies with data [1].

There are also other ethical features of using AI in journalism. News channels must provide a way to describe their usage of AI in their content. Many concerns are raised while AI maintains strong biases in the data provided. The automatic way of collecting content for news has a threat that most of the journalists lose their careers. One major concern is quality and correctness of the data provided by the AI content. AI's evolution will help news organizations identify more creative ways to make use of the technology. Integrating AI into journalism works best when the strengths of AI are measured by the ethical and main values of journalism.

4 AI in News Production and Distribution

4.1 Automated Video and Audio Editing

AI has improved how multimedia content files are developed by automating a lot of tedious tasks. These tools use machine learning algorithms to help improve workflow which makes editing videos, transcribing audio and teamwork more efficient. AI can check through clips, issue first cuts, adjust colors, smooth out shaky shots and remove specific objects from the video. Audio transcripts benefit from improved accuracy when speech recognition algorithms are used and AI tools also measure audience engagement, suggest how to organize content, recommend music and create custom thumbnails [20]. Using AI tools on the cloud helps different team members to collaborate more efficiently. While AI makes manufacturing more efficient, it helps people, instead of replacing them in tasks that require creativity and choice. As technology improves, AI will likely be used more to automate many stages of multimedia production such as AI storyboards, automatic translation and support for writing scripts.

4.2 Chatbots and Conversational AI in News Delivery

More news websites are using AI-driven chatbots to involve their audience and show them relevant information. Chatbots with these features use natural language processing and machine learning to understand users, what they like to read and give personalized updates. They provide several services, for example, making personalized news feeds, sending instant updates, managing user questions, helping find content, speaking many languages, being available at all hours, learning what readers prefer, improving resources, offering up-to-date forms of news highlights and allowing news gatherings by voice. Yet, there are problems and ethical considerations associated with using AI chatbots in news such as making sure information is accurate, lowering the chance of bias, remaining transparent, looking after user privacy, preventing people from being surrounded by opinions and making sure a human supervises the process. The development of AI will likely allow chatbots to deliver more news which calls on news organizations to use both technology and traditional journalism together.

4.3 Predictive Analytics for Content Strategy

With historical and current data, AI algorithms make it much easier to estimate what the audience will look for in content. They depend on machine learning to find trends and links in the actions of their audiences. They study seasonality, ongoing developments and cultural behavior to forecast changes in the market. They use insights to set strategy and change their methods, choosing the most ideal types of content [21]. New data leads to machines learning continuously and their performance improves over successive applications. The planning of advertising content relies heavily on a person's ability to weigh relevant factors and brand attributes against one another, and even when AI is used to support that process, it's an artist's brain that's doing the calculations.

4.4 AI-Enhanced Search and Archiving Systems

Modern AI has transformed how news archives are managed and accessed, assisting journalism workflows operated at a quicker speed. Humans have a long way to go in understanding and filtering all the news content to find what they need faster, and they use machine learning algorithms and technology to filter and analyze the variety of news content. AI systems capable of accurately analyzing and classifying content do so through natural language processing algorithms that help the system process human language. The result is that news articles are automatically tagged with keywords, topics and entities, providing clear organization that allows for easy retrieval of documents later. Unlike keyword matching, these search engines can grasp the background and intent behind the search queries due to the use of AI [22]. Researchers and journalists have an easy task finding needed information as well as understanding the material, although in some cases there are no matching keywords included.

Through machine learning algorithms, news articles are linked based on similar patterns and the evolution of certain subjects can be examined as well as less obvious connections among stories [23]. This skill makes it possible for journalists to explain complicated matters well and spot new viewpoints. By using AI, researchers and reporters can speed up their work since they can see main ideas from a mass of information through content summarization. The summaries provide reporters with the ability to follow a subject or to briefly find the latest news. News article creation interests, the changes that occurred to the popular sentiment, and indicators of media biasness are tracked with the assistance of algorithms in both the short term and over time. Using this information, they are able to analyse the way in which a news topic is being discussed and the way in which the sentiments of the people shift across a period of time.

5 Ethical Considerations and Challenges in AI-Powered Journalism

The issue of AI in journalism is associated with the emergence of ethical issues and challenges that need to be addressed. With the rise in the usage of AI in the newsrooms, organizations need to actively review the ways these systems can raise the risk already in place.

5.1 Algorithmic Bias and Fairness

Will AI-generated journalism mirror what was available in its training data? Since AI learns according to what it has observed, it might recreate all obscure inclinations in the information it views or in the architecture of the framework. This may cause disbalance in the news coverage. There is a risk that in generating an AI-based recommendation engine, the information will be driven into a few political perspectives and there is no sense of equal meetings and balanced reporting [24]. News companies ought to thoroughly test their AI engines to make sure their systems are functioning in the right manner. Such steps involve the utilization of training data of different types, consistent bias audits and engagement of multidisciplinary teams. Journalists and editors should be conscious of some kind of existing bias in case they can lessen it in AI-aided content.

5.2 Transparency and Accountability in AI-Generated Content

As AI presents daily financial reports, as well as personal news, it becomes difficult to recognize human writing, rather than machine one. As a result, the responsibility and clarity of journalism is doubted by many people. The news outlets have to make available easily readable policies concerning the reporting of matters relating to AI as well as policies pertaining to editorial matters. In this respect this implies that readers should also be unaware when information is being produced or modified by the AI and how AI is being utilized in the production and distribution of content. Openness develops faith and challenges the listeners. It is also necessary that there is accountability. Companies need to establish processes of how to handle AI-composed stories, allocating human workers to spot and fix AI mistakes, and have quick mechanisms available to fix AI-generated errors.

5.3 Privacy and Data Protection Concerns

The algorithms in journalism representing AI typically begins with gathering masses of user-data in order to provide personal news. Although this makes it more involving and relevant, it poses serious privacy and data-security concerns. News organizations should also strike a balance with how personal data is handled with the obligation to fix them. Journalists must have very strong security systems and user explications as to why their information is gathered. Also, some ethical dilemmas are encountered when the information in data-driven journalism exposes sensitive personal data or when attempting to market it. Following regulations such as general data protection, whether required by law or not, is a moral as well as a legal requirement [25]. To ensure trust and standard ethics in journalism, organizations should surpass standard requirements and follow high standards of data and user privacy protection.

5.4 Job Displacement and the Changing Role of Human Journalists

There are worries that AI applications will cut down on jobs for journalists. Although they can handle data analysis, transcription and news-writing, AI is more likely to change journalists' jobs than to replace them entirely. Since AI can do some of the work, human journalists will have to focus on skills that need human thought and emotion in their tasks. Journalists should be given new training programs by both news businesses and schools to help them succeed with AI in newsrooms. They should be able to use new tools, think critically, make principled choices and operate AI systems with the same dedication to good journalism.

6 The Future of AI in Journalism

6.1 Augmented Reality and Immersive News Experiences

AR tools powered by AI can change how news is experienced by readers. Interactive three-dimensional experiences made possible by these technologies greatly improve the way people grasp and interact with news. Being able to effortlessly explore subjects by using AR is one advantage of AI in news storytelling. For example, using AR in a news story about climate change, people could see how sea levels could affect a spot nearby. This kind of diagram makes it possible to see and grasp abstract concepts.

Data gathered by AI is used to produce responsive real-time displays that reflect what a person wants or where they are [26]. An AR application for news could show economic information that is relevant to each user or the place they live in.

Customizing the news in this way ensures the stories are meaningful and important to every possible user. Intelligent features make it possible for members of the audience to become actively involved. Those using the story could play with the 3D models, view different possibilities or supply their own data. Interactivity allows users to engage more with the subject and learn more, as they interact and experiment in the AR platform.

Graphics from augmented reality (AR), including translations, captions or audio descriptions, can be added to news stories using AI to make them more accessible. Due to this approach, many more people can participate in and gain from engaging with immersive news stories. By combining AI and AR, art can adapt the way it tells a story to match a user's reactions and how they interact with it [27]. AI can check a viewer's expressions, look or reactions to decide their level of involvement and adjust the content shown to them.

In addition, AR made possible with AI can encourage groups of people to engage with news together. Several users can look at the same AR environment and discuss their thoughts at the same time. News consumed with others can open discussions and make learning more engaging for all. Algorithms can offer help with verifying facts and giving extra context to what you see in the AR experience. While users are reading a news story, AI can mention background information, confirm facts in real time and supply links to extra reports on the topic.

Data visualization holds great promise in AR news content. Interactive three-dimensional models can be built from large data using computer tools [28]. With this approach, complex financial reports, scientific research or statistics can be made simpler for most readers. These tools can also help news become more understanding by allowing people to feel what it's like to be in someone else's shoes through AR. A story that looks at refugees might make use of AR to create scenarios for readers, connecting them more strongly and helping them perceive the issue clearly.

In essence, new information delivery strategies are made possible using AI in augmented reality for news stories. Making news experiences interactive and three-dimensional can increase how much people learn, get engaged and keep news information which benefits society by fostering understanding and caring.

6.2 AI-Assisted Multilingual Journalism

World journalism could be changed greatly when advanced language models enable the quick translation and localization of current news [29]. This new technology can help break down language issues, so information can be distributed to more people easily. With these models, responding to crisis or urgent news situations happens quickly, so important information reaches people quickly, regardless of language. Because they are so affordable, fewer people are needed to do the translations. They can keep terms and styles the same throughout the translation process, which ensures that all media reports are the same. Besides, modern models can be improved for certain topics which makes them more accurate in journalism.

But there are some difficulties when using these language models in journalism. There is still worry about the accuracy of models since they usually miss fine details that are closely related to a specific topic. The ethical issue of transparency, accountability and fairness is important regarding the use of automated processes. Since automatic translations don't always catch culture-specific details, they might cause confusion. AI translation may affect how people earn their living as professional translators and interpreters [30]. But the targeted and excessive use of computers could become a danger to the quality and precision of news dissemination. A better alternative is to control automation alongside human efforts to close the gap in delivery. However, all new technologies might help to connect the media of different parts of the world and give more understanding and opportunity to learn about new information. Before using these technologies, journalists should check ethical guidelines, implement quality checks which they can trust and follow up with the results. News organizations must balance the new access and efficiency while not compromising on good journalism, or their culture. Advanced language models have the potential to bring more equality and cross-language communication to the world of journalism.

6.3 Predictive Journalism and Scenario Modeling

Because AI algorithms can process a vast amount of historical and current data, they can identify patterns, trends and correlations that may be missed by unaided analysts. AI is used to examine historical information and provide news businesses with recurring patterns and forecasts of what will happen again in the future, thereby enabling them to plan accordingly. Here you will find stories of the seasons, political actions and economic trends repeatedly. In providing an analysis of current events, artificial intelligence (AI) provides news companies with an understanding of how events might impact the future, which in turn influences how they are covered and the resources they dedicate to coverage [31]. Such algorithms allow news organizations to be alerted to issues that are beginning to gain traction, so that they are the first to cover such news. Using reader engagement data, AI experts can learn the interests of most people and adjust articles and stories to keep pace with them. AI can assist news organizations by analyzing geopolitical information in order to anticipate future issues, changes in diplomacy and economic shifts.

The continuously updated narration about climate change, technology or society should change to be effective. They need a strategy that adheres to changing events and takes advantage of opportunities, and it is why AI becomes indispensable. Predictive analytics allow organizations to determine where they need to cover by the time accidents happen. Governed by AI algorithms, journalists are able to examine large volumes of available data simultaneously, extracting the connections between trends that appear unrelated, but may suggest new content and broader reporting. Though the AI facilitates planning of news, humans should not be left out, and ethics are to be respected. Data miners and journalists must easily consider the AI data, and even better surely no matter how famous that the algorithms release does endings up impacting human judgment, reasonable journalism attitudes, or the discovery of stories that are generally ignored. The use of AI in news planning should support, not eliminate, the main rules of journalism, helping to make better choices and decisions in the newsroom [32].

6.4 Human-AI Collaboration in Newsrooms

Bringing AI into journalism can greatly improve the news industry by enhancing several parts of how news is made, delivered and used. AI enhances the process of data collection and analysis, accelerates template stories generation, verifies facts, and tailors content to the reader, and distributed content efficiently. These profits liberate human journalists to focus on explaining events, writing detailed stories and carrying out investigations, work that do not depend on technology to accomplish but rather on emotional understanding and ethical decision making.

A close collaboration between journalists and AI explains why the former can leave AI with regular assignments free to work on higher-value tasks. But such integration will cost newsrooms in terms of investment in new technology, additional training and worries about job loss. Open conversations on how AI will be applied, as well as explicit instructions, will keep the population trustful [33]. With the help of AI, journalism can survive online and maintain its fundamental principles and values.

7 Preparing for an AI-Driven Journalism Landscape

In order to remain up to date with the ways journalism is swiftly evolving, it is essential to introduce AI and data science. Technology has bigger roles in reporting news, research and communication. AI is transforming the industry through the possibility of auto writing, personalizing news, and effective data estimation. Nowadays, journalists need to learn about AI algorithms, about possible biases and reflect on ethical concerns. They must think discretely about AI-generated stories and be responsible with the use of AI tools in reporting. Data science skills are also essential. As big data is everywhere, journalists must have the ability to gather, analyze, and obtain meaningful insights of large datasets to find something new [34]. It also requires a knowledge of the data visualization approaches to make complex things clear.

Topics in curriculum integration should include an exploration of the basic concepts of AI, fundamentals of machine learning, understanding data analysis and statistical methods, starting with programming languages such as Python and R, working with data visualization methods, taking into consideration the implication of AI in ethics and data usage and using AI to curate and produce different types of content [35]. Students get a good understanding of AI approach which illustrates how AI is applied to many fields. AI is best used in task-focused projects such as reporting news, accessing people's opinions via social media or presenting stories that are data-rooted. The skills for teamwork with data professionals and specialists in AI are important for students since it is common in current newsrooms.

Journalism organizations need to consider the ethical and practical issues of AI in journalism. It covers the issue of privacy, discriminatory algorithms, and AI potential to journalists. These courses are important to enable students to be comfortable with new technology, as that way, they can apply AI and data science in reporting, storytelling, and analysis. Through perfect use of these tools, journalists would be able to keep beside of the latest ways of communication and at the same time they will be able to keep in mind the main tenets of journalism.

7.1 Adapting Newsroom Workflows and Organizational Structures

To make the best use of AI, news organizations have to redefine their processes completely, should journalists wish to utilize AI. They should have a roadmap such that they can be balanced in new technology, ethics and good journalism. Instead of implementing AI, it is important to analyze existing workflows and find out where AI can be useful. These activities involve article production, data, analyzing, fact-checking, and interacting with the audience. These processes have to be mapped to get organizations a picture of where AI can be utilized.

News organizations should then aid their staff in learning about AI. They need to know about AI features, difficulties and what might be considered unethical, adds the report. If everyone in an organization is aware of how technology works, they can use AI effectively following guidelines from journalism [36]. It is best to add AI technologies gradually and do so several times. News businesses can start using AI in areas where it would help such as automatically tagging news and offering personalized suggestions to readers. As employees are trained in these tools, AI can start being used for difficult processes like computer-generated routine news articles or AI support during investigative work.

It is very important for news organizations to put in place clear rules and guidelines about AI. They should handle matters like algorithmic bias, how clear AI content is and why it still needs human supervision. Having strict review steps protects the accuracy, fairness and accountability in the AI-generated content. Moreover, it would be helpful for news organizations to either set up ethics committees regarding AI or hire AI ethics officers who oversee the daily and continued use of AI by the company. The groups should be able to provide input on ethical arguments, review AI decisions and ensure that the organization's use of AI aligns with its own journalistic mission and values.

AI can be successfully applied when development teams and editing teams work together. News organizations should promote interdisciplinary collaboration between data scientists, AI professionals, journalists and editors. This collaboration will ensure that the purposes and the standards of AI used in journalism are met. With the advancement of technology, news organizations should remain flexible and soft to new developments. AI software and the impact it has on journalism should be reviewed on a regular basis [37]. Changes to how a company operates should be considered as well as changes to ethical standards and potential action to remove or update artificial intelligence tools that conflict with company values.

Furthermore, it is crucial for news organizations to engage in discussions and initiatives that unite the industry in the utilization of AI. In the cases when the industry gathers tips, fixes obstacles and pays attention to ethical norms, it will be able to collaborate to establish AI utilization norms. Media business owners need to consider AI as an augmentation of journalism and no longer as a substitute. The philosophy in question ought to be incorporated into the processes to demonstrate that AI is a human tool, as it will keep creativity and critical thinking. Thoughtful evolutions will allow organizations to apply AI to make operations more rapid, correct and scaled by observing the journalistic standards. This strategy will enable media to flourish in contemporary times and keep society informed.

7.2 Establishing Ethical Guidelines and Best Practices

It is crucial to introduce ethical ground rules governing the use of AI in journalism to ensure that they are not abused and that the profession stays ethical. They need to address the way data is used, reduction of bias, the way algorithms are explained, check facts, etc. To ensure that AI-assisted journalism is honest and well done, journalists should have rules about checking sources, labeling articles and appropriate training. Journalists, technologists and ethicists must develop systems of accountability and collaboration to find solutions in the field. Adaptive governance, the active involvement of people and protocols in case of a crisis should be part of the standards. Diversity and inclusion in media, standards of quality and intellectual property are all important [38]. Some standards can only be effective across borders if they deal with issues faced cross-border, and whistleblower protections are only effective cross-border if they protect whistleblowers across borders. Journalism is headed towards responsible AI by creating AI audit tools and by regularly evaluating AI's outputs. Compliance with these standards can enable the industry to use AI effectively and retain its core values and its public trust.

7.3 Fostering Interdisciplinary Collaboration

Collaboration between journalists, data scientists and researchers is one way forward for AI to advance journalism [39]. While journalists can bring helpful skills in how to tell stories, how to make ethical decisions, and how to communicate with people, data scientists contribute their data analysis tools and approaches in visualizing computer data. AI researchers are familiar with machine learning and natural language processing which enables them to create tools for automatic tasks and better content creation [40]. Several problems of AI-based news coverage became feasible: better news research, trusted fact-checking, tailored feeds, robots and interesting data visualizations. Ai also has the ability to fill language portals, to be used responsibly, delve into predictive journalism, to increase visitor interest, and generate new forms of business. News outlets are able to fix difficult issues by coming up with creative solutions. In turn, these partnerships can be built in educational institutions and media organizations via interdisciplinary work, hackathons, and research events. In the face of AI technologies, such collaborations would become crucial to the future of journalism and would work together to use the assets to rise to the occasion and get opportunities.

8 Conclusion

Implementation of artificial intelligence in the journalism sector represents a shift of paradigm never experienced in the history of introduction of technology into journalism. It compels a review of ways of doing things, frameworks, and democracy. This chapter illustrates the disruptive effect of AI in the areas of operational efficiency, ethical governance, and democratic preservation. It has been demonstrated that routine workflows in the newsroom, better debt of processing and fact-checking as well as content customization are already automated using machine learning, natural language processing, and computer vision. But automated misinformation and algorithm bias are endangering editorial integrity and require immediate systems intervention. Digital media production forecasts the fall of previous revenue patterns, which necessitates hybrid monetise plans like paywalls, sponsored content, and targeted advertising, and there is a total redefinition of the connection among journalism and its audience. The conclusions made by those findings imply that strong AI application should entail a high level of journalism, algorithm benchmarking, bias arising, and techno aptitude. The future of journalism is not either on human versus artificial intelligence but about using it in harmony to serve democratic society.

Declaration of Generative AI and AI-Assisted Technologies in the Chapter Preparation Process During the preparation of this work the authors used Perplexity in order to review the flow of chapter. After using this tool/service, the authors reviewed and edited the content as needed and take full responsibility for the content of the published article.

References

1. Kevin-Alerechi, E., et al.: AI and the newsroom: transforming journalism with intelligent systems. J. Artif. Intell. Mach. Learn. Data Sci. **3**(1), 1930–1937 (2025)
2. Verma, D.: Impact of artificial intelligence on journalism: a comprehensive review of AI in journalism. J. Commun. Manage. **3**(02), 150–156 (2024)
3. Ozurumba, C.J., Fagbohun, O, Akintola, A.S.: Ethically integrating AI into journalism: balancing innovation, integrity, and public trust. Proceedings of ISETE International Conference, Istanbul, Turkey, 30-36 (2024)
4. Alzub, A.M.: Navigating the disruption of digital and conventional media in changing media consumption landscape in digital era. J. Eng. Technol. Appl. Sci. (JETAS) **5**(1), 38–48 (2023)
5. Gupta, S.S.: Digital revolution: journalism in the internet age. Hist. J.Ism. **65** (2023)
6. O'Brien, D.: 30 The transformation of news in the digital age: business model changes, challenges, and future directions. In: Rohn, U., Bjørn von Rimscha, M., Raats, T. (eds.) De Gruyter Handbook of Media Economics, pp. 431–450. De Gruyter (2024)
7. Audry, S.R., Navila, I.: Adaptive strategies of digital transformation on students' consumptive behavior in the era of society 5.0. Cirebon Int. J. Econ. Bus. **1**(2), 171–179 (2023)
8. Asimakopoulos, G., et al.: Impact of information and communication technologies on democratic processes and citizen participation. Societies **15**(2), 40 (2025)
9. de-Lima-Santos, M.F., et al.: Digital news business models in the age of industry 4.0: digital Brazilian news players find in technology new ways to bring revenue and competitive advantage. Digit. J. **12**(9), 1304–1328 (2024)
10. Oeldorf-Hirsch, A., Srinivasan, P.: An unavoidable convenience: how post-millennials engage with the news that finds them on social and mobile media. Journalism **23**(9), 1939–1954 (2022)
11. Westlund, O., Krumsvik, A.H., Lewis, S.C.: Competition, change, and coordination and collaboration: tracing news executives' perceptions about participation in media innovation. J. Stud. **22**(1), 1–21 (2021)
12. Trattner, C., et al.: Responsible media technology and AI: challenges and research directions. AI Ethics **2**, 585–594 (2022)
13. Das, D., Upadhyay, A.K.: Examining the emergence of new journalistic forms and innovation in the Indian newspaper industry. Hum. Soc. Sci. Commun. **11**(1), 1–12 (2024)
14. Pradeep, E., et al.: Empowering social media users with ethical AI. In: Ethical AI Solutions for Addressing Social Media Influence and Hate Speech, pp. 91–118. IGI Global Scientific Publishing (2025)
15. Ünver, A.: Emerging technologies and automated fact-checking: tools, techniques and algorithms (2023)
16. Hassani, H., et al.: Artificial intelligence (AI) or intelligence augmentation (IA): what is the future? AI **1**(2), 8 (2020)
17. Leelavathy, S., Nithya, M.: Public opinion mining using natural language processing technique for improvisation towards smart city. Int. J. Speech Technol. **24**(3), 561–569 (2021)
18. Habil, S., El-Deeb, S., El-Bassiouny, N.: AI-based recommendation systems: the ultimate solution for market prediction and targeting. In: The Palgrave Handbook of Interactive Marketing, pp. 683–704. Springer International Publishing, Cham (2023)
19. Fu, Y., Stasko, J.: More than data stories: broadening the role of visualization in contemporary journalism. IEEE Trans. Vis. Comput. Graph. **30**(8), 5240–5259 (2023)
20. Eftekhari, H.: Transcribing in the digital age: qualitative research practice utilizing intelligent speech recognition technology. Eur. J. Cardiovasc. Nurs. **23**(5), 553–560 (2024)
21. Nixon, L., et al.: AI and data-driven media analysis of TV content for optimised digital content marketing. Multimed. Syst. **30**(1), 25 (2024)
22. Subramanian, D.V., et al.: KMSBOT: enhancing educational institutions with an AI-powered semantic search engine and graph database. Soft. Comput. **29**(1), 1–15 (2025)
23. Gallofré Ocaña, M., Opdahl, A.L.: Supporting newsrooms with journalistic knowledge graph platforms: current state and future directions. Technologies **10**(3), 68 (2022)

24. Leiser, M.R.: Bias, journalistic endeavours, and the risks of artificial intelligence. In: Artificial Intelligence and the Media, pp. 8–32. Edward Elgar Publishing (2022)
25. Labadie, C., Legner, C.: Building data management capabilities to address data protection regulations: learnings from EU-GDPR. J. Inf. Technol. **38**(1), 16–44 (2023)
26. Ouyang, W.: Data visualization in big data analysis: applications and future trends. J. Comput. Commun. **12**(11), 76–85 (2024)
27. Nikolarakis, A., Koutsabasis, P.: Mobile AR interaction design patterns for storytelling in cultural heritage: a systematic review. Multimodal Technol. Interact. **8**(6), 52 (2024)
28. Schachner, E.R., et al.: Perspectives on lung visualization: three-dimensional anatomical modeling of computed and micro-computed tomographic data in comparative evolutionary morphology and medicine with applications for COVID-19. Anat. Rec. **308**(4), 1118–1143 (2025)
29. Lin, J.: Artificial intelligence applications in English news translation: strategies and research. In: International Conference on Educational Information Technology, vol. 6, no. 2 (2024)
30. Agunlejika, T.: Exploring the potential of AI in news translation for rural communities in Nigeria: towards a conceptual framework. Available SSRN. 5132798 (2025)
31. Kieslich, K., Diakopoulos, N., Helberger, N.: Anticipating impacts: using large-scale scenario-writing to explore diverse implications of generative AI in the news environment. AI Ethics **5**(5), 4555–4577 (2025)
32. Sonni, A.F., et al.: Digital newsroom transformation: a systematic review of the impact of artificial intelligence on journalistic practices, news narratives, and ethical challenges. J.Ism Media. **5**(4), 1554–1570 (2024)
33. Spyridou, P., Ioannou, M.: Exploring AI amid the hype: a critical reflection around the applications and implications of AI in journalism. Societies **15**(2), 23 (2025)
34. Vinay, S.: Data scientist competencies and skill assessment: a comprehensive framework. J. ID. **1660**, 1544 (2024)
35. Hsu, Y.-C.: Mapping the landscape of data science education in higher general education in Taiwan: a comprehensive syllabi analysis. Educ. Sci. **14**(7), 763 (2024)
36. Murire, O.T.: Artificial intelligence and its role in shaping organizational work practices and culture. Adm. Sci. **14**(12), 316 (2024)
37. Calvo-Rubio, L.-M., Rojas-Torrijos, J.-L.: Criteria for journalistic quality in the use of artificial intelligence. Commun. Soc. **37**(2), 247–259 (2024)
38. Kerrigan, P., Liddy, S., O'Brien, A.: Tools and measures for diversity and inclusion in media industries: international best practice and informing policy change in the Irish film and television sector. Eur. J. Commun. **38**(3), 217–236 (2023)
39. de-Lima-Santos, M.F.: The entanglements between data journalism, collaboration and business models: a systematic literature review. Digit. J. **12**(2), 256–281 (2024)
40. Alqahtani, T., et al.: The emergent role of artificial intelligence, natural learning processing, and large language models in higher education and research. Res. Soc. Adm. Pharm. **19**(8), 1236–1242 (2023)

The manufacturer's authorised representative in the EU is Springer Nature Customer Service Centre GmbH, Europaplatz 3, 69115 Heidelberg, Germany. If you have any concerns regarding our products, please contact ProductSafety@springernature.com

Printed and bound by CPI Group (UK) Ltd, Croydon, CR0 4YY
21/09/2026
02219992-0005